DISCLAIMER

TABLE OF CONTENTS

LIST OF TABLES

LIST OF FIGURES

CHAPTER 1: OVERVIEW

1.1 INTRODUCTION

The Department of Energy's (DOE) Carbon Capture program is conducted under the Clean Coal Research Program (CCRP). DOE's overarching mission is to increase the energy independence of the United States and to advance U.S. national and economic security. To that end, the DOE Office of Fossil Energy (FE) has been charged with ensuring the availability of ultraclean (near-zero emissions), abundant, low-cost domestic energy from coal to fuel economic prosperity, strengthen energy independence, and enhance environmental quality. As a component of that effort, the CCRP—administered by the Office of Clean Coal and implemented by the National Energy Technology Laboratory (NETL)—is engaged in research, development, and demonstration (RD&D) activities to create technology and technology-based policy options for public benefit. The CCRP is designed to remove environmental concerns related to coal use by developing a portfolio of innovative technologies, including those for carbon capture and storage (CCS). The CCRP comprises two major program areas: CCS and Power Systems and CCS Demonstrations. The CCS and Power Systems program area is described in more detail below. The CCS Demonstrations program area includes three key subprograms: Clean Coal Power Initiative, FutureGen 2.0, and Industrial Carbon Capture and Storage. The technology advancements resulting from the CCS and Power Systems program area are complemented by the CCS Demonstrations program area, which provides a platform to demonstrate advanced coal-based power generation and industrial technologies at commercial scale through cost-shared partnerships between the Government and industry.

While it has always been an influential component of CCS research, recently DOE has increased its focus on carbon utilization to reflect the growing importance of developing beneficial uses for carbon dioxide (CO_2). At this time, the most significant utilization opportunity for CO_2 is in enhanced oil recovery (EOR) operations. The CO_2 captured from power plants or other large industrial facilities can be injected into existing oil reservoirs. The injected CO_2 helps to dramatically increase the productivity of previously depleted wells—creating jobs, reducing America's foreign oil imports, and thus increasing energy independence. Simultaneously, the CO_2 generated from power production is stored permanently and safely. The CCRP is gathering the data, building the knowledge base, and developing the advanced technology platforms needed to prove that CCS can be a viable strategy for reducing greenhouse gas (GHG) emissions to the atmosphere, thus ensuring that coal remains available to power a sustainable economy. Program efforts have positioned the United States as the global leader in clean coal technologies.

This document serves as a program plan for NETL's Carbon Capture research and development (R&D) effort, which is conducted under the CCRP's CCS and Power Systems program area. The program plan describes the Carbon Capture R&D efforts in 2013 and beyond. Program planning is a strategic process that helps an organization envision the future; build on known needs and capabilities; create a shared understanding of program challenges, risks, and potential benefits; and develop strategies to overcome the challenges and risks, and realize the benefits. The result of this process is a technology program plan that identifies performance targets, milestones for meeting these targets, and a technology pathway to optimize R&D activities. The relationship of the Carbon Capture program to the CCS and Power Systems program area is described in the next section.

1.2 CCS AND POWER SYSTEMS PROGRAM AREA

The CCS and Power Systems program area conducts and supports long-term, high-risk R&D to significantly reduce fossil fuel power-plant emissions (including CO_2) and substantially improve efficiency, leading to viable, near-zero-emissions fossil fuel energy systems. The success of NETL research and related program activities will enable CCS technologies to overcome economic, social, and technical challenges including cost-effective CO_2 capture, compression, transport, and storage through successful CCS integration with power-generation systems; effective CO_2 monitoring and verification; permanence of underground CO_2 storage; and public acceptance. The overall program consists of four subprograms: Advanced Energy Systems, Carbon Capture, Carbon Storage, and Crosscutting Research (Figure 1-1). These four subprograms are further divided into numerous Technology Areas. In several instances, the individual Technology Areas are further subdivided into key technologies.

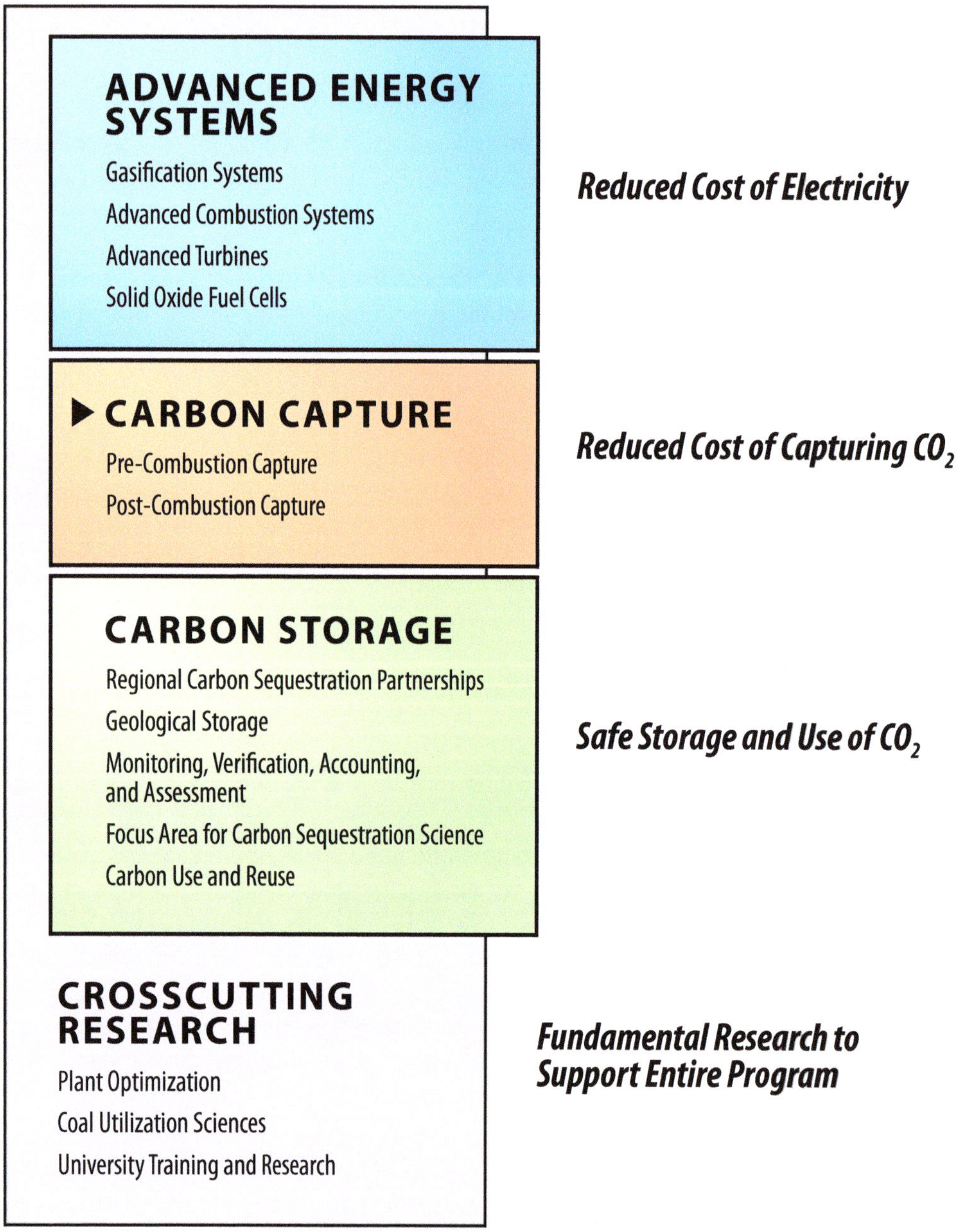

Figure 1-1. CCS and Power Systems Subprograms

The ***Advanced Energy Systems subprogram*** is developing a new generation of clean fossil fuel-based power systems capable of producing affordable electric power while significantly reducing CO_2 emissions. This new generation of technologies will essentially be able to overcome potential environmental barriers and meet any projected environmental emission standards. A key aspect of the Advanced Energy Systems subprogram is targeted at improving overall thermal efficiency, including the capture system, which will be reflected in affordable CO_2 capture and reduced cost of electricity (COE). The Advanced Energy Systems subprogram consists of four Technology Areas as described below:

- ***Gasification Systems*** research to convert coal into clean high-hydrogen synthesis gas (syngas) that can in-turn be converted into electricity with over 90 percent CCS.
- ***Advanced Combustion Systems*** research that is focused on new high-temperature materials and the continued development of oxy-combustion technologies.
- ***Advanced Turbines*** research, focused on developing advanced technology for the integral electricity-generating component for both gasification and advanced combustion-based clean energy plants fueled with coal by providing advanced hydrogen-fueled turbines, supercritical CO_2-based power cycles and advanced steam turbines.
- ***Solid Oxide Fuel Cells*** research is focused on developing low-cost, highly efficient solid oxide fuel cell power systems that are capable of simultaneously producing electric power from coal with carbon capture when integrated with coal gasification.

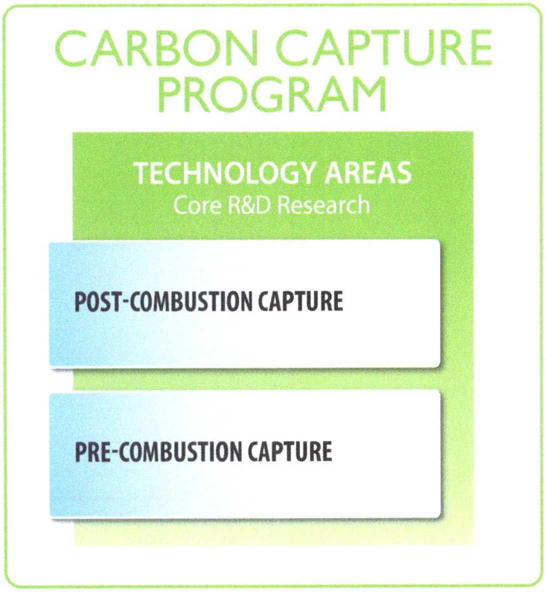

Figure 1-2. Carbon Capture Program Technology Areas

The ***Carbon Capture subprogram*** is focused on the development of post-combustion and pre-combustion CO_2 capture technologies for new and existing power plants (Figure 1-2). Post-combustion CO_2 capture technology is applicable to conventional combustion-based power plants, while pre-combustion CO_2 capture is applicable to gasification-based systems. In both cases, R&D is underway to develop solvent-, sorbent-, and membrane-based capture technologies.

The ***Carbon Storage subprogram*** advances safe, cost-effective, permanent geologic storage of CO_2. The technologies developed and large-volume injection tests conducted through this subprogram will be used to benefit the existing and future fleet of fossil fuel power-generating facilities by developing tools to increase our understanding of geologic reservoirs appropriate for CO_2 storage and the behavior of CO_2 in the subsurface.

The ***Crosscutting Research subprogram*** serves as a bridge between basic and applied research by fostering the R&D of instrumentation, sensors, and controls targeted at enhancing the availability and reducing the costs of advanced power systems. This subprogram also develops computation, simulation, and modeling tools focused on optimizing plant design and shortening developmental timelines, as well as other crosscutting issues, including plant optimization technologies, environmental and technical/economic analyses, coal technology export, and integrated program support.

The CCS and Power Systems program area is pursuing three categories of CCS and related technologies referred to as 1st-Generation, 2nd-Generation, and Transformational. These categories are defined in Figure 1-3.

1st-Generation Technologies—include technology components that are being demonstrated or that are commercially available.

2nd-Generation Technologies—include technology components currently in R&D that will be ready for demonstration in the 2020–2025 timeframe.

Transformational Technologies—include technology components that are in the early stage of development or are conceptual that offer the potential for improvements in cost and performance beyond those expected from 2nd-Generation technologies. The development and scaleup of these "Transformational" technologies are expected to occur in the 2016–2030 timeframe, and demonstration projects are expected to be initiated in the 2030–2035 time period.

Figure 1-3. CCS Technology Category Definitions

1.3 THE RD&D PROCESS

The research, development, and demonstration of advanced fossil fuel power-generation technologies follows a sequential progression of steps toward making the technology available for commercial deployment, from early analytic study through pre-commercial demonstration. Planning the RD&D includes estimating when funding opportunity announcements (FOAs) will be required, assessing the progress of ongoing projects, and estimating the costs to determine budget requirements.

1.3.1 TECHNOLOGY READINESS LEVELS

The Technology Readiness Level (TRL) concept was adopted by the National Aeronautics and Space Administration (NASA) to help guide the RD&D process. TRLs provide an assessment of technology development progress on the path to meet the final performance specifications. The typical technology development process spans multiple years and incrementally increases scale and system integration until final-scale testing is successfully completed. The TRL methodology is defined as a "systematic metric/measurement system that supports assessments of the maturity of a particular technology and the consistent comparison of maturity between different types of technology."[1] Appendix A includes a table of TRLs as defined by DOE's Office of Fossil Energy.

The TRL score for a technology is established based upon the scale, degree of system integration, and test environment in which the technology has been successfully demonstrated. Figure 1-4 provides a schematic outlining the relationship of those characteristics to the nine TRLs.

1 Mankins, J., Technology Readiness Level White Paper, 1995, rev. 2004, Accessed September 2010. http://www.artemisinnovation.com/images/TRL_White_Paper_2004-Edited.pdf

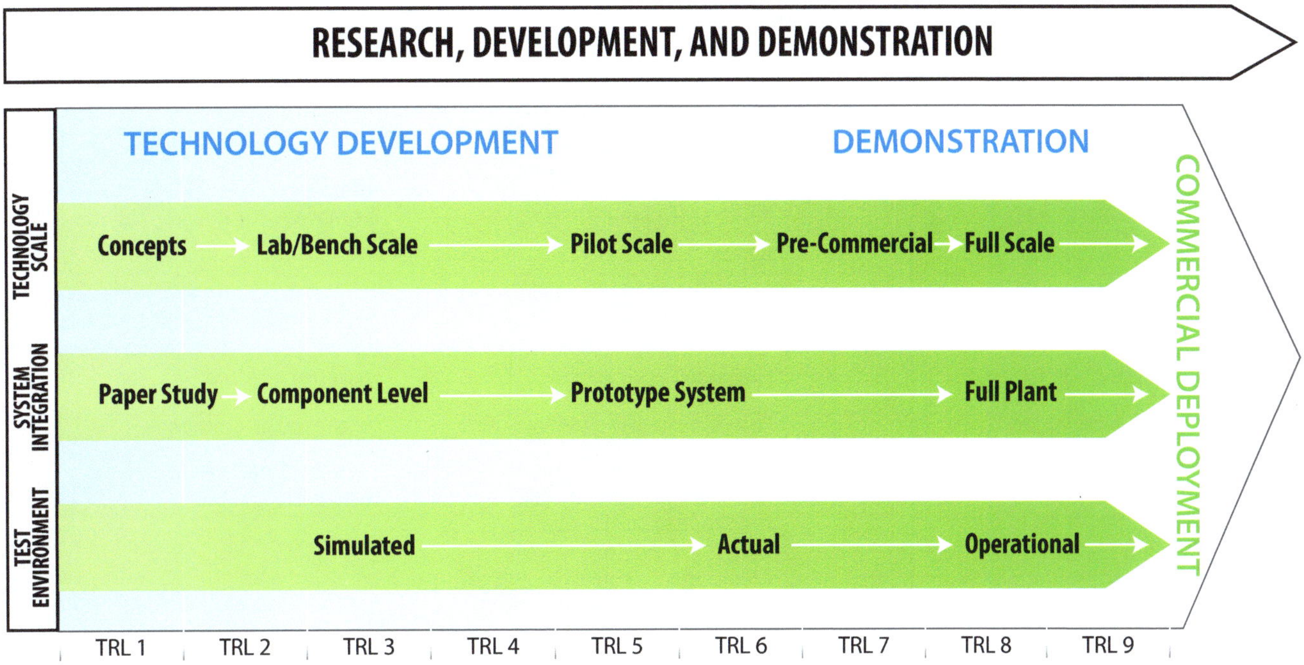

Figure 1-4. Technology Readiness Level—Relationship to Scale, Degree of Integration, and Test Environment

The scale of a technology is the size of the system relative to the final scale of the application, which in this case is a full-scale commercial power-production facility. As RD&D progresses, the scale of the tests increases incrementally from lab/bench scale, to pilot scale, to pre-commercial scale, to full-commercial scale. The degree of system integration considers the scope of the technology under development within a particular research effort. Early research is performed on components of the final system, a prototype system integrates multiple components for testing, and a demonstration test of the technology is fully integrated into a plant environment. The test environment considers the nature of the inputs and outputs to any component or system under development. At small scales in a laboratory setting it is necessary to be able to simulate a relevant test environment by using simulated heat and materials streams, such as simulated flue gas or electric heaters. As RD&D progresses in scale and system integration, it is necessary to move from simulated inputs and outputs to the actual environment (e.g., actual flue gas, actual syngas, and actual heat integration) to validate the technology. At full scale and full plant integration, the test environment must also include the full range of operational conditions (e.g., startup and turndown).

1.3.2 RD&D RISK AND COST PROGRESSION

As the test scale increases, the duration and cost of the projects increase, but the probability of technical success also tends to increase. Given the high technical risk at smaller scales, there will often be several similar projects that are simultaneously supported by the program. On the other hand, due to cost considerations, the largest projects are typically limited to one or two that are best-in-class. Figure 1-5 provides an overview of the scope of laboratory/bench-, pilot-, and demonstration-scale testing in terms of test length, cost, risk, and test conditions. In the TRL construct, "applied research" is considered to be equivalent to lab/bench-scale testing, "development" is carried out via pilot-scale field testing, and "large-scale testing" is the equivalent of demonstration-scale testing. The CCS and Power Systems program area encompasses the lab/bench-scale and pilot-scale field testing stages and readies the technologies for demonstration-scale testing.

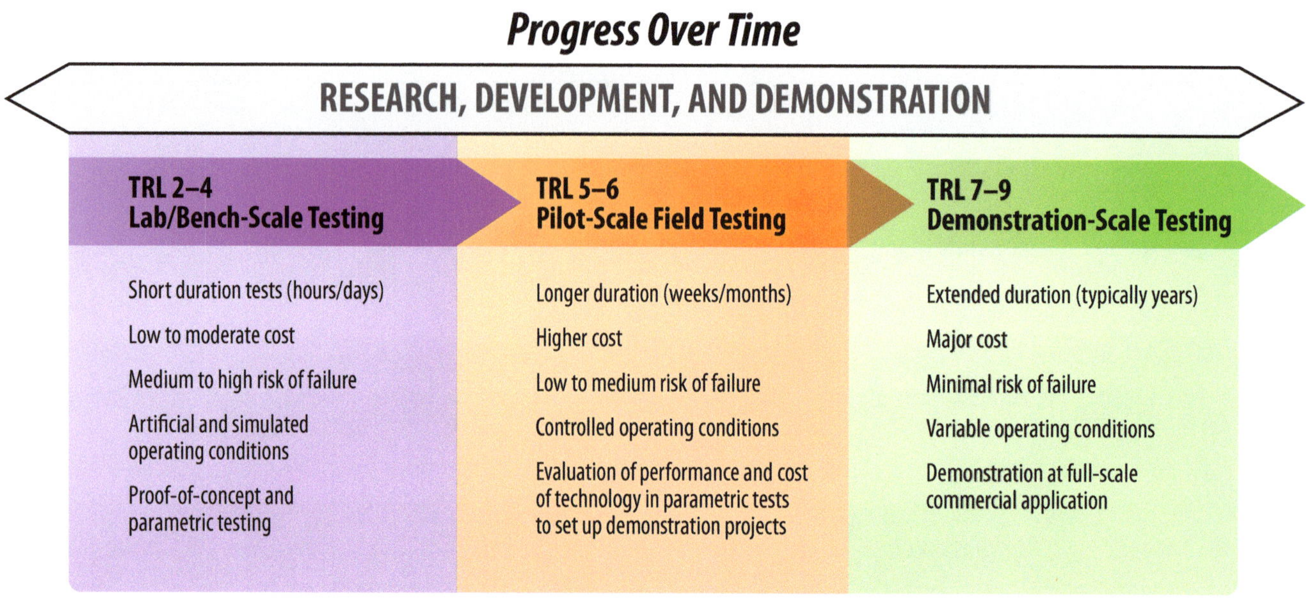

Progress Over Time
RESEARCH, DEVELOPMENT, AND DEMONSTRATION
TRL 2–4
Lab/Bench-Scale Testing
Short duration tests (hours/days)
Low to moderate cost
Medium to high risk of failure
Artificial and simulated operating conditions
Proof-of-concept and parametric testing
TRL 5–6
Pilot-Scale Field Testing
Longer duration (weeks/months)
Higher cost
Low to medium risk of failure
Controlled operating conditions
Evaluation of performance and cost of technology in parametric tests to set up demonstration projects
TRL 7–9
Demonstration-Scale Testing
Extended duration (typically years)
Major cost
Minimal risk of failure
Variable operating conditions
Demonstration at full-scale commercial application

Figure 1-5. Summary of Characteristics at Different Development Scales

1.4 BARRIERS/RISKS AND MITIGATION STRATEGIES

The risk and mitigation strategy to achieving all performance targets by 2030 is summarized in Table 1-1. The overarching challenge to be addressed by Carbon Capture is to economically generate clean energy using fossil fuels. The same barriers, risks, and mitigation strategies apply to all Technology Areas.

Table 1-1. **Issues/Barriers and Mitigation Strategies**

Issue	Barrier/Risk	Mitigation Strategy
Cost: Economically generating clean energy using fossil fuels Performance: Achieve performance targets by 2030 Environment: Meet near-zero emissions (including >90% CO_2 capture) with minimal cost impact Market: Low economic growth; natural gas price Regulations: Uncertainties	Existing/new plants do not adopt advanced Carbon Capture technologies Lower natural gas prices Reduced Carbon Capture program budget	Near-, mid-, and long-term R&D projects to foster the commercialization of advanced technologies Comprehensive, multipronged R&D approach to advanced CO_2 capture technologies

THIS PAGE INTENTIONALLY LEFT BLANK

CHAPTER 2: CARBON CAPTURE PROGRAM

2.1 INTRODUCTION

The Carbon Capture program consists of two core research Technology Areas: (1) Post-Combustion Capture and (2) Pre-Combustion Capture. Post-combustion capture is primarily applicable to conventional pulverized coal (PC)-fired power plants, where the fuel is burned with air in a boiler to produce steam that drives a turbine/generator to produce electricity. The carbon is captured from the flue gas after fuel combustion. Pre-combustion capture is applicable to integrated gasification combined cycle (IGCC) power plants, where solid fuel is converted into gaseous components (syngas) by applying heat under pressure in the presence of steam and oxygen. In this case, the carbon is captured from the syngas before combustion and power production occurs. Although R&D efforts are focused on capturing CO_2 from the flue gas or syngas of coal-based power plants, the same capture technologies are applicable to natural gas- and oil-fired power plants and other industrial CO_2 sources.

Current R&D efforts conducted within the Carbon Capture program include development of advanced solvents, sorbents, and membranes for both the Post- and Pre-Combustion Technology Areas (Figure 2-1). The research focus for these post-combustion and pre-combustion technologies is presented in Chapter 4. Under both Technology Areas, the program is developing 2nd-Generation and Transformational CO_2 capture technologies that have the potential to provide step-change reductions in both cost and energy penalty as compared to currently available 1st-Generation technologies. The success of the program in developing these technologies will enable cost-effective implementation of CCS throughout the power-generation sector and ensure that the United States will continue to have access to safe, reliable, and affordable energy from fossil fuels.

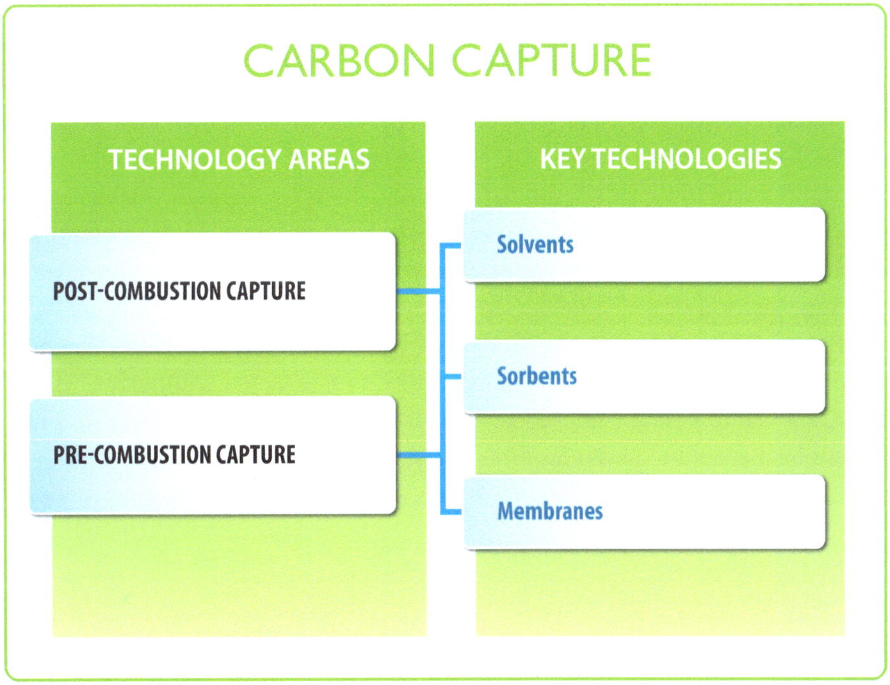

Figure 2-1. Technology Areas and Key Technologies for Carbon Capture Program

2.2 BACKGROUND

CCS begins with the separation and capture of CO_2 from coal-based power plant flue gas or syngas. There are commercially available 1st-Generation CO_2 capture technologies that are currently being used in various industrial applications. However, in their current state of development these technologies are not ready for implementation on coal-based power plants because they have not been demonstrated at appropriate scale, require approximately one-third of the plant's steam and power to operate, and are very expensive. For example, DOE/NETL estimates that the deployment of a current 1st-Generation post-combustion CO_2 capture technology—chemical absorption with an aqueous monoethanolamine solution—on a new PC power plant would increase the COE by ≈80 percent and derate the plant's net generating capacity by as much as 30 percent. Other major challenges include energy integration, flue gas contaminants, water use, CO_2 compression, and oxygen supply for pre-combustion systems.

The net electrical output from a coal-based power plant employing currently available 1st-Generation CO_2 capture and compression technologies will be significantly less than that for the same plant without capture. This is because some of the energy—thermal and electrical—produced at the plant must be used to operate the CO_2 capture and compression processes. Steam usage decreases the gross electrical generation, while the additional auxiliary power usage decreases the net electrical output of the power plant. Implementation of CO_2 capture results in a 7–10 percentage point decrease in net plant efficiency depending on the type of power-generation facility.

The energy penalty associated with CO_2 capture has been estimated in a DOE study conducted to determine the cost and performance of a post-combustion CO_2 capture technology retrofit on American Electric Power's coal-fired Conesville Unit No. 5. The amine-based CO_2 capture process would require extraction of approximately 50 percent of the steam that normally flows through the low-pressure turbine for a 90 percent CO_2 capture scenario. Consequently, the gross power output of the unit would decrease by ≈16 percent (from 463.5 MWe to 388.0 MWe). In addition, the auxiliary power requirements for the CO_2 capture and compression system would be 55 MWe. The combined effect of steam and auxiliary power required to operate the CO_2 capture and compression system is a reduction in the net power output of the unit by approximately 30 percent (from 433.8 MWe to 303.3 MWe).

Post-combustion CO_2 capture is primarily applicable to conventional coal-, oil-, or gas-fired power plants, but could also be applicable to IGCC and natural gas combined cycle flue gas capture. A simplified block diagram illustrating the post-combustion CO_2 capture process is shown in Figure 2-2. In a typical coal-fired power plant, fuel is burned with air in a boiler to produce steam that drives a turbine/generator to produce electricity. Flue gas from the boiler consists mostly of nitrogen (N_2) and CO_2. The CO_2 capture process would be located downstream of the conventional pollutant controls for nitrogen oxides (NO_x), particulate matter (PM), and sulfur dioxide (SO_2). Chemical solvent-based technologies currently used in industrial applications are being considered for this purpose. The chemical solvent process requires the extraction of a relatively large volume of low-pressure steam from the power plant's steam cycle, which decreases the gross electrical generation of the plant. The steam is required for release of the captured CO_2 and regeneration of the solvent. Separating CO_2 from this flue gas is challenging for several reasons: a high volume of gas must be treated (≈2 million cubic feet per minute for a 550-MWe plant), the CO_2 is dilute (between 12 and 14 percent CO_2), the flue gas is at atmospheric pressure, trace impurities (PM, SO_2, NO_x, etc.) can degrade capture media, and compressing captured CO_2 from near-atmospheric pressure to pipeline pressure (about 2,200 pounds per square inch absolute [psia]) requires a large auxiliary power load.

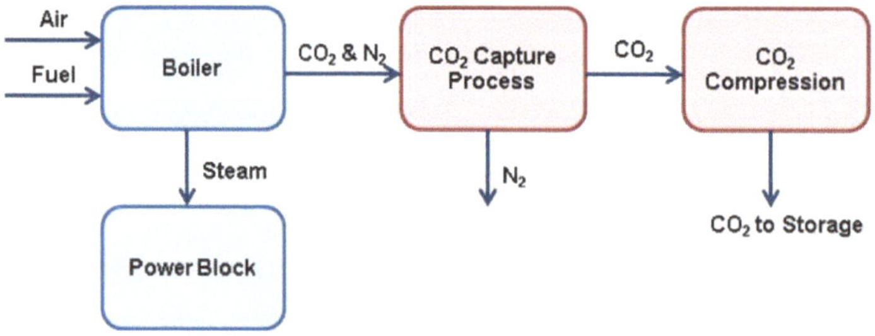

Figure 2-2. Block Diagram Illustrating Power Plant with Post-Combustion CO_2 Capture

Pre-combustion capture is mainly applicable to gasification plants, where fuel (coal, biomass, or coal/biomass mixture) is converted into gaseous components by applying heat under pressure in the presence of steam and sub-stoichiometric oxygen (O_2). A simplified block diagram illustrating the pre-combustion CO_2 capture process is shown in Figure 2-3. By carefully controlling the amount of O_2, only a portion of the fuel burns to provide the heat necessary to decompose the fuel and produce syngas, a mixture of hydrogen (H_2) and carbon monoxide, and minor amounts of other gaseous constituents. To enable pre-combustion capture, the syngas is further processed in a water-gas-shift (WGS) reactor, which converts carbon monoxide into CO_2 while producing additional H_2, thus increasing the CO_2 and H_2 concentrations. An acid-gas-removal system can then be used to separate the CO_2 from the H_2. Physical solvent-based technologies currently used in industrial applications are being considered for this purpose. After CO_2 removal, the H_2-rich syngas is used as a fuel in a combustion turbine combined cycle to generate electricity.

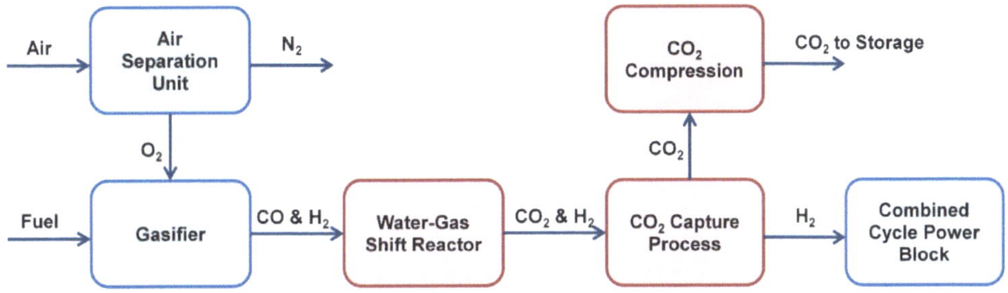

Figure 2-3. Block Diagram Illustrating Power Plant with Pre-Combustion CO_2 Capture

2.3 RECENT R&D ACTIVITIES

Current R&D efforts conducted within the Carbon Capture program are focused on development of advanced solvents, sorbents, and membranes for both the Post-Combustion and Pre-Combustion Technology Areas. Research projects are carried out using various funding mechanisms—including partnerships, cooperative agreements, and financial assistance grants—with corporations, small businesses, universities, nonprofit organizations, and other national laboratories and Government agencies. Current efforts cover the development of 2nd-Generation and Transformational CO_2 capture technologies. Although the majority of these technologies are still in the laboratory- and bench-scale stages of development, a limited number of small pilot-scale field tests have been initiated. Table 2-1 presents the number of active projects per Technology Area and test scale. A complete list of active Carbon Capture projects is presented in Appendix B.

Table 2-1. **Active CO_2 Capture Technology R&D Projects**

CO_2 Capture Technology Pathway	Laboratory/Bench-Scale Projects	Small Pilot-Scale Projects
Post-Combustion	28	6
Pre-Combustion	8	0

More details on the specific advanced capture technologies currently under development are available in the report entitled, *DOE/NETL Advanced Carbon Dioxide Capture R&D Program: Technology Update*, which is available at: http://www.netl.doe.gov/technologies/coalpower/ewr/pubs/CO2Handbook/

2.4 IMPORTANT ASPECTS OF CARBON CAPTURE PROGRAM R&D PROCESS

The Carbon Capture program comprises a comprehensive effort to develop cost-effective, advanced, post-, and pre-combustion technologies for power plants and other industrial facilities that significantly reduce the energy penalty and capital cost compared to currently available 1st-Generation technologies. The RD&D process to develop these technologies includes several important aspects, including (1) putting together pieces of the technology puzzle, (2) progress over time, and (3) technology down-selection and scaleup.

2.4.1 PUTTING TOGETHER THE PIECES OF THE PUZZLE

The development of a 2nd-Generation or Transformational CO_2 capture technology includes more than laboratory-scale testing of process chemistry and physics and evaluation of associated operating parameters. The research effort can also involve the development of new chemical production methods, novel process equipment designs, new equipment manufacturing methods, and optimization of the process integration with other power-plant systems (e.g., the steam cycle, cooling water system, and CO_2 compression system). Figure 2-4 presents the various R&D components that might be necessary to take a capture technology from concept to commercial reality. Developing a successful CO_2 capture technology requires putting together all these pieces of the puzzle. While some of these developments are unique to a specific process, others could be more generally applicable. For example, a novel process equipment design developed by one research organization could prove vital to optimizing performance of the process chemistry developed by another research organization. While most of the CO_2 capture technology projects encompass the entire range of R&D components, there are some that focus more on a specific component or perhaps are more successful with a specific component (e.g., process chemistry or process equipment design). As a result, it could take the integration of the successful development of multiple components from multiple researchers to eventually arrive at a successful and cost-effective CO_2 capture technology. For example, a post-combustion, solvent-based CO_2 capture process might require a synthesis of the following "pieces" to be judged an overall technology success: (1) an advanced solvent with superior working capacity and regeneration energy requirements, (2) an advanced absorption reactor with improved mass transfer capability, and (3) an advanced regeneration reactor/re-boiler that minimizes energy requirements. The successful development of these three separate technology "pieces" could rely on three separate projects, rather than a single project.

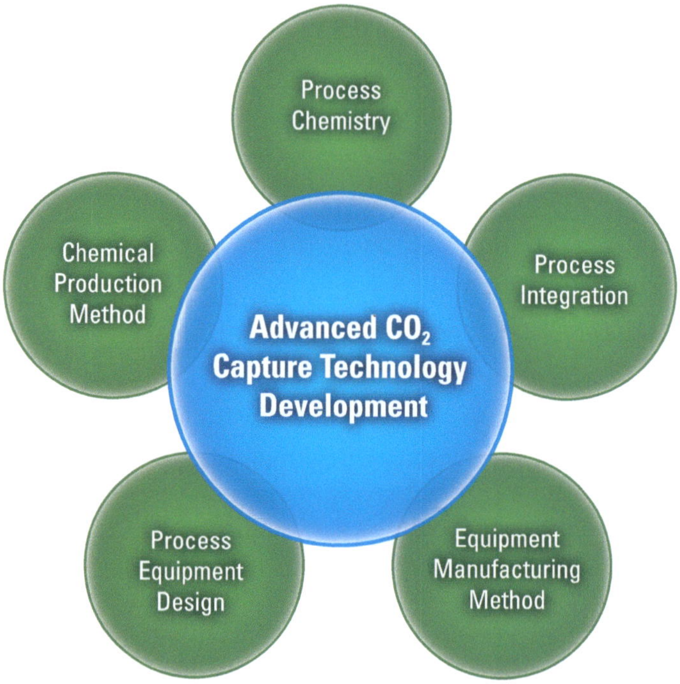

Figure 2-4. Components of CO_2 Capture Technology Development

2.4.2 PROGRESS OVER TIME

DOE/NETL envisions having a 2nd-Generation CO_2 capture technology portfolio ready for demonstration-scale testing after 2020 following the sequential progression of laboratory-, bench-, and pilot-scale testing. Similarly, a Transformational CO_2 capture technology portfolio should be ready for demonstration-scale testing after 2030.

As noted previously with regard to the R&D process—generally, there is a relatively high risk of failure associated with laboratory/bench-scale testing, a lower risk of failure for pilot-scale testing, and a minimal risk of failure for full-scale demonstrations. Specifically with regard to CO_2 capture technology development, laboratory- and bench-scale testing is usually conducted with simulated flue or synthesis gas at relatively low gas flow rates ranging from 1 to 100 standard cubic feet per minute (scfm). Small pilot-scale testing can also be conducted in a laboratory setting as a "semi-batch" mode using coal combustors to generate flue gas for process testing. For example, the University of North Dakota's Energy and Environmental Research Center (UNDEERC) uses two sizes of combustors for small pilot-scale testing with equivalent gas flow rates of approximately 10 scfm and 125 scfm.

Upon completion of laboratory- and bench-scale testing, it is necessary to conduct pilot-scale slipstream testing using actual flue gas to determine potential adverse effects on the process from minor constituents in the coal that are present in the syngas or combustion flue gas. For example, trace concentrations of arsenic in some coals were found to poison the catalyst used in the SCR process for control of NO_x from coal-fired power plants. Likewise, low concentrations of SO_2 are known to degrade amine solvent performance. In addition, potential problems with excessive scaling, plugging, and/or corrosion of process equipment can be evaluated and solutions developed only via operating experience during long-term, pilot-scale slipstream testing.

The flue gas design flow rates for NETL's large pilot-scale slipstream testing, including those conducted at the National Carbon Capture Center (NCCC), will be in the range of 1,000–12,000 scfm. For comparison, a 1-MW gross electric-generation facility produces approximately 2,500 scfm of combustion flue gas. After successful completion of pilot-scale testing, the process equipment can be further scaled up to conduct demonstration-scale testing prior to commercial deployment of the technology.

PROGRESS OVER TIME

An example of the scaleup process is the RD&D being conducted by Membrane Technology and Research, Inc. (MTR) to develop a new membrane-based post-combustion CO_2 capture technology. MTR and NETL initiated the membrane R&D program in 2007. MTR's first phase of R&D included bench-scale testing of various membrane designs using a simulated gas flow rate of approximately 2.5 scfm. Based on successful bench-scale testing, MTR initiated a follow-up project with NETL in 2008 to conduct a small slipstream field test that was conducted in 2010. The approximately 175-scfm (equivalent to approximately 1 ton per day [tpd] of CO_2) testing was conducted at the Arizona Public Service's coal-fired Cholla Power Plant located in Arizona. MTR now plans to conduct additional small pilot-scale field testing based on a gas flow rate of approximately 2,500 scfm (equivalent to approximately 1 MWe or about 20 tpd of CO_2) as part of a new project with NETL that is scheduled for completion in 2015.

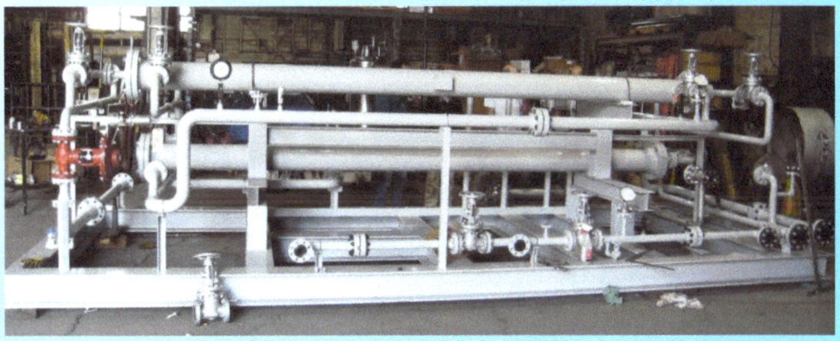

2.4.3 TECHNOLOGY DOWN-SELECTION AND SCALEUP

As described previously, a sequential progression of scaleup testing is necessary to accommodate the RD&D process. As shown previously in Figure 1-5, there is a relatively high risk of failure associated with laboratory/bench-scale testing, a lower risk of failure for pilot-scale testing, and a minimal risk of failure for full-scale demonstrations. Therefore, in order to ensure a reasonable overall probability of success in eventually developing 2nd-Generation and Transformational CO_2 capture technologies under the program, it is necessary to start with a relatively large portfolio of laboratory/bench-scale projects. Figure 2-5 presents this concept as an RD&D development "funnel" that portrays the down-selection and corresponding scaleup process.

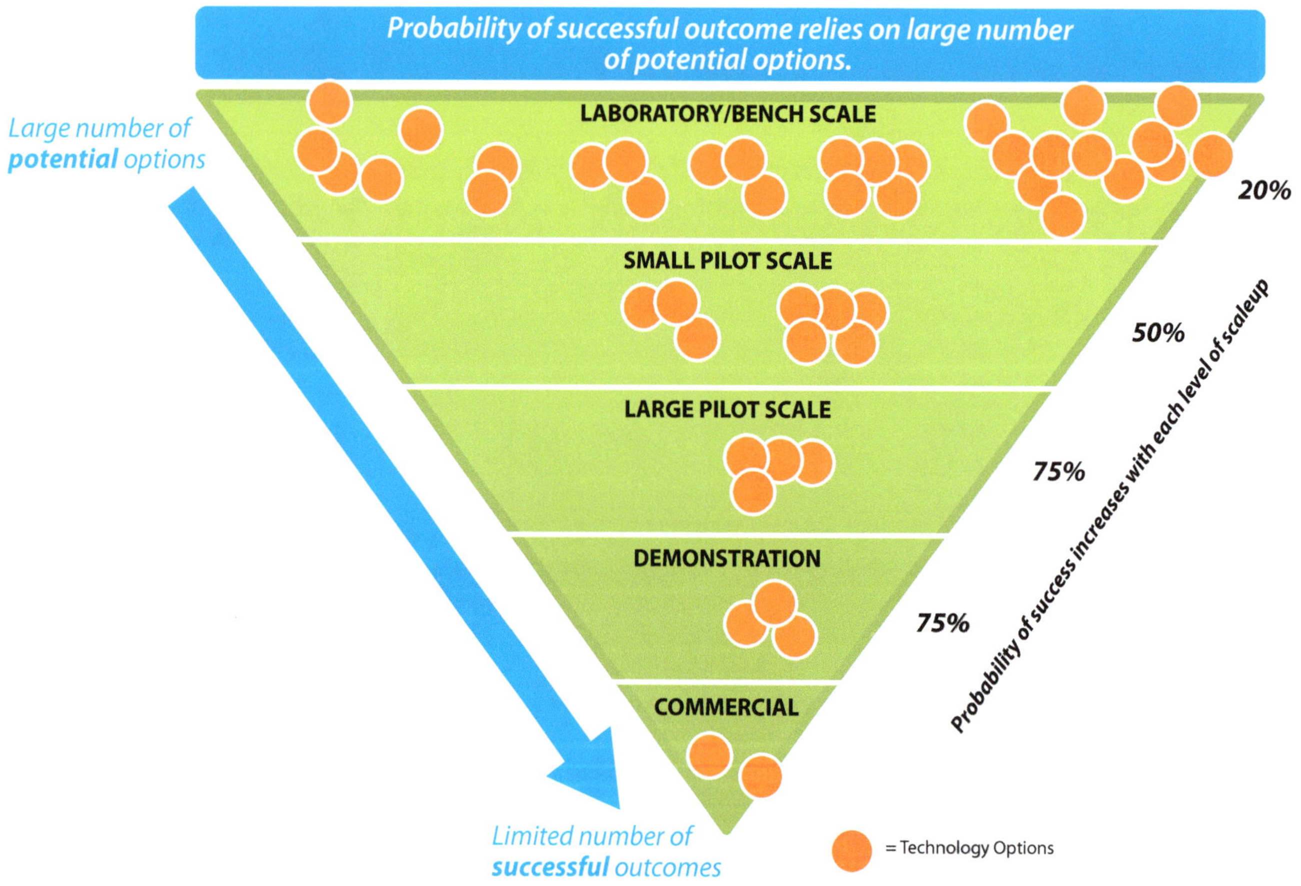

Figure 2-5. CO_2 Capture Technology RD&D Funnel

In this funnel example, a successful outcome for the Carbon Capture R&D program is the development of two commercially available CO_2 capture technologies after a four-step down-selection process that progresses from a large portfolio of laboratory/bench-scale projects with high risk/low probability of success to a small portfolio of full-scale demonstration projects with low risk/high probability of success. The progressive down-selection process accompanying the technology scaleup is necessary because there is no practical means to predict eventual commercial success based on laboratory/bench-scale test results.

THIS PAGE INTENTIONALLY LEFT BLANK

CHAPTER 3: GOALS AND BENEFITS

3.1 GOALS

The goals of the Carbon Capture program support the energy goals established by the Administration, DOE, FE, and the CCRP. The priorities, mission, goals, and targets of each of these entities are summarized in Appendix C.

3.1.1 CCRP GOALS

Currently, the CCRP is pursuing the demonstration of 1st-Generation CCS technologies with existing and new power plants and industrial facilities using a range of capture alternatives and storing CO_2 in a variety of geologic formations. In parallel, to drive down the costs of implementing CCS, the CCRP is pursuing RD&D to decrease the COE and capture costs and increase base power-plant efficiency, thereby reducing the amount of CO_2 that has to be captured and stored per unit of electricity generated. FE is developing a portfolio of technology options to enable this country to continue to benefit from using our secure and affordable coal resources. The challenge is to help position the economy to remain competitive, while reducing carbon emissions.

There are a number of technical and economic challenges that must be overcome before cost-effective CCS technologies can be implemented. The experience gained from the sponsored demonstration projects focused on state-of-the-art (1st-Generation) CCS systems and technologies will be a critical step toward advancing the technical, economic, and environmental performance of 2nd-Generation and Transformational systems and technologies for future deployment. In addition, the core RD&D projects being pursued by the CCRP leverage public and private partnerships to support the goal of broad, cost-effective CCS deployment. The following long-term performance goals for new coal-fired power generation facilities have been established for the CCRP (alternate goals have been established for retrofit applications, as discussed in the next section):

- Develop 2nd-Generation technologies that:
  - Are ready for demonstration in the 2020–2025 timeframe (with commercial deployment beginning in 2025)
  - Cost less than \$40/tonne of CO_2 captured
- Develop Transformational technologies that:
  - Are ready for demonstration in the 2030–2035 timeframe (with commercial deployment beginning in 2035)
  - Cost less than \$10/tonne of CO_2 captured

The planning necessary to implement the above goals and targets is well underway and the pace of activities is increasing. The path ahead with respect to advancing CCS technologies, particularly at scale, is very challenging given today's economic risk-averse climate and that no regulatory framework is envisioned in the near term for supporting carbon management. These conditions have caused DOE/FE to explore a strategy with increased focus on carbon utilization as a means of reducing financial risk. This strategy benefits from FE's investment in the beneficial utilization of CO_2 for commercial purposes, particularly through the development of next-generation CO_2 injection/EOR technology, with the objective of creating jobs and increasing energy independence. Carbon dioxide injection/EOR is a specific market-based utilization strategy that will positively impact domestic oil production and economical CO_2 capture and storage.

3.1.2 CARBON CAPTURE PROGRAM GOALS

CCRP cost and performance goals, summarized in Table 3-1, can be met by an integrated system if the underlying technology components are successfully developed. The Carbon Capture program supports achievement of the CCRP goals by developing advanced, efficient carbon capture technologies that produce ultraclean (near-zero emissions, including CO_2), low-cost energy with low water use. In support of those overall goals, specific cost and performance goals for 2025 and 2035 are described below.

Table 3-1. **Market-Based R&D Goals for Advanced Coal Power Systems**

	Goals (for nth-of-a-kind plants)		Performance Combinations that Meet Goals	
R&D Portfolio Pathway	*Cost of Captured CO_2, $/tonne*[1]	*COE Reduction*[2]	*Efficiency (HHV)*	*Capital/O&M Reduction*[3]
2nd-Geneneration R&D Goals for Commercial Deployment of Coal Power in 2025[4]				
In 2025, EOR revenues will be required for 2nd-Generation coal power to compete with natural gas combined cycle and nuclear in absence of a regulation-based cost for carbon emissions.				
Greenfield Advanced Ultra-Supercritical PC with CCS	40	20%	37%	13%
Greenfield Oxy-Combustion PC with CCS	40	20%	35%	18%
Greenfield Advanced IGCC with CCS	≤40	≥20%	40%	18%
Retrofit of Existing PC with CCS	45	n/a		
Transformational R&D Goals for Commercial Deployment of Coal Power in 2035[4]				
Beyond 2035, Transformational R&D and a regulation-based cost for carbon emissions will enable coal power to compete with natural gas combined cycle and nuclear without EOR revenues.				
New Plant with CCS—Higher Efficiency Path	<10[5]	40%	56%	0%
New Plant with CCS—Lower Cost Path	<10[5]	40%	43%	27%
Retrofit of Existing PC with CCS	30	≥40%	n/a	
Transformational pathways could feature advanced gasifiers, advanced CO_2 capture, 3,100 °F gas turbines, supercritical CO_2 cycles, pulse combustion, direct power extraction, pressurized oxy-combustion, chemical looping, and solid oxide fuel cells.				

NOTES:

(1) Assumes 90 percent carbon capture. First-year costs expressed in 2011 dollars, including compression to 2,215 psia but excluding CO_2 transport and storage (T&S) costs. The listed values do not reflect a cost for carbon emissions, which would make them lower. For greenfield (new) plants, the cost is relative to a 2nd-Generation ultra-supercritical PC plant without carbon capture. For comparison, the nth-of-a-kind cost of capturing CO_2 from today's IGCC plant, compared to today's supercritical PC without carbon capture, is about $60/tonne. For retrofits, the cost is relative to the existing plant without capture, represented here as a 2011 state-of-the-art subcritical PC plant with flue gas desulfurization and selective catalytic reduction. The cost of capturing CO_2 via retrofits will vary widely based on the characteristics of the existing plant such as its capacity, heat rate, and emissions control equipment. The nth-of-a-kind cost of capture for retrofitting the representative PC plant described above (a favorable retrofit target) using today's CO_2 capture technology would be about $60/tonne. (In contrast, today's first-of-a-kind cost of CO_2 capture for a new or existing coal plant is estimated to be $100–$140/tonne.)

(2) Relative to the first-year COE of today's state-of-the-art IGCC plant with 90 percent carbon capture operating on bituminous coal, which is currently estimated at $133/MWh. For comparison, the first-year COE of today's supercritical PC with carbon capture is estimated to be $137/MWh. Values are expressed in 2011 dollars. They include compression to 2,215 psia but exclude CO_2 T&S costs and CO_2 EOR revenues. However, CO_2 T&S costs were considered, as appropriate, when competing against other power-generation options in the market-based goals analysis.

(3) Cost reduction is relative to today's IGCC with carbon capture. Total reduction is comprised of reductions in capital charges, fixed operating and maintenance (O&M) and non-fuel variable O&M costs per million British thermal unit (Btu) (higher heating value [HHV]) of fuel input. Cost reductions accrue from lower equipment and operational costs, availability improvements, and a transition from high-risk to conventional financing. The ability to secure a conventional finance structure is assumed to result from lowering technical risk via commercial demonstrations.

(4) 2nd-Generation technologies will be ready for large-scale testing in 2020, leading to commercial deployment by 2025 and attainment of nth-of-a-kind performance consistent with R&D goals by 2030. Transformational technologies will be ready for large-scale testing in 2030, leading to initial commercial deployment in 2035 and attainment of nth-of-a-kind performance consistent with R&D goals by 2040.

(5) Cost of captured CO_2 ranges from $5 to $7/tonne for the cost reductions and efficiencies noted.

For IGCC operation, the pre-combustion capture technology employed is one part of the overall system that will lead to achievement of 2nd-Generation and Transformational cost of capture goals. As illustrated in Figure 3-1, cost reductions for IGCC systems are achieved through advances in technologies being pursued by the Advanced Turbines, Gasification Systems, and Crosscutting Research technology programs. The goal for development of a 2nd-Generation pre-combustion capture system is a $5/tonne reduction in capture cost. The goal for Transformational pre-combustion capture technologies is an additional reduction in capture cost, beyond 2nd-Generation, of $5/tonne.

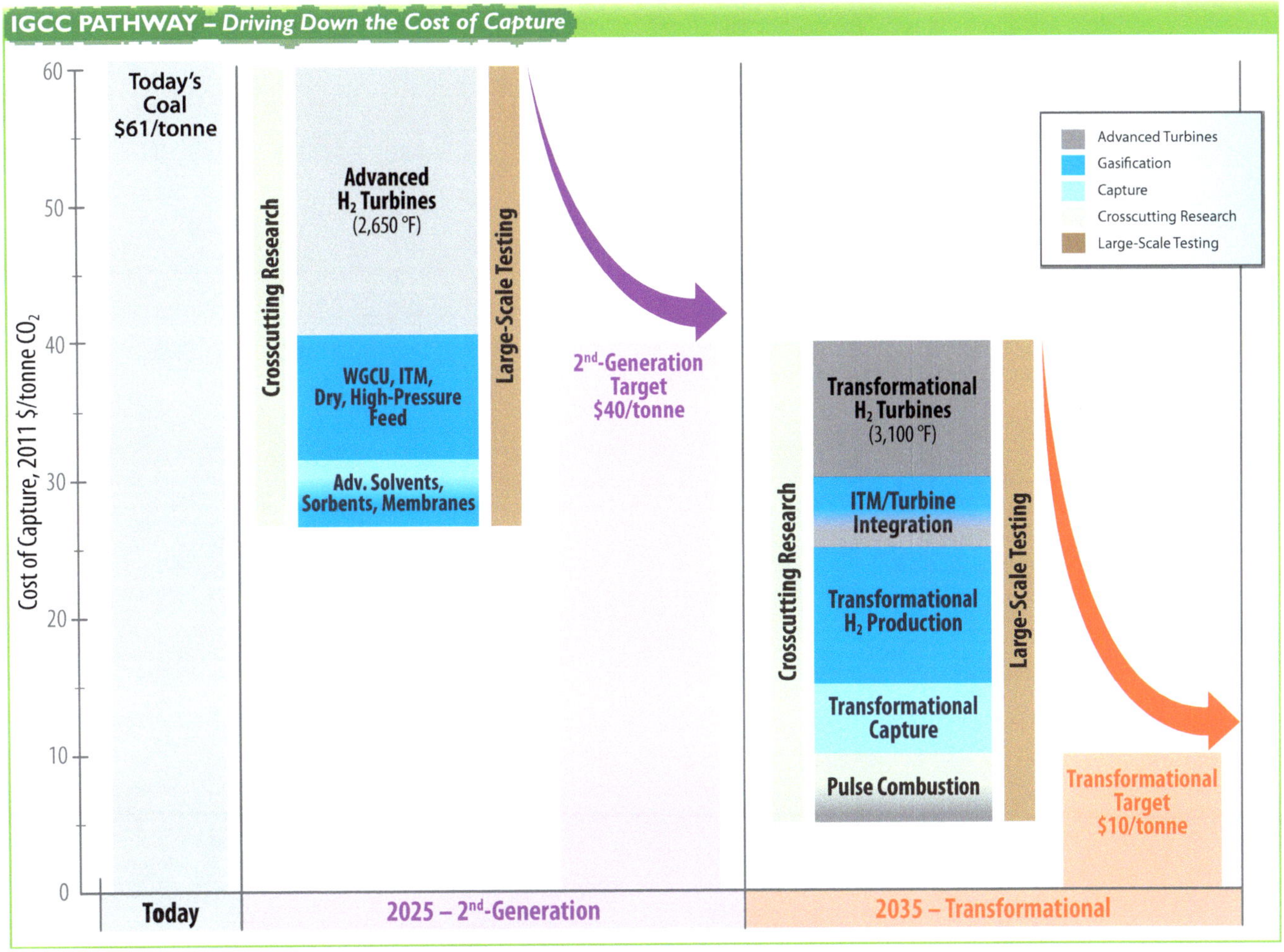

Figure 3-1. Targets for Technology Contributions to Overall CCRP Cost of Capture Goals—IGCC Pathway

Establishing program goals for post-combustion capture requires a slightly different approach because the technologies are applicable to both new plants and existing plant retrofits. For new plants, 2nd-Generation and Transformational capture cost goals of $40/tonne and <$10/tonne are targeted to be met through a combination of technologies in the Carbon Capture R&D and Crosscutting Research programs and the Advanced Turbines Technology Area, as illustrated in Figure 3-2. Approximately $12/tonne of the capture cost reduction is targeted to result from implementation of 2nd-Generation post-combustion CO_2 capture and compression technologies. Transformational advancements in CO_2 capture technologies, including improved power-plant integration designs and opportunities for cosequestration of other criteria pollutants are targeted to reduce the cost of capture an additional $22/tonne (from $40/tonne to $18/tonne) by 2030. For both 2nd-Generation and Transformational technologies, the remaining capture cost reductions required to meet CCRP goals will come from advancements in technologies under development within different programs and Technology Areas. For example, a Transformational advanced capture solution ready for testing in 2030 may consist of the following:

- New capture material (solid or liquid) + advanced unit operation + advanced integrated process + advanced CO_2 compression
- Multipollutant capture and/or cosequestration ($Hg + SO_x + CO_2$)

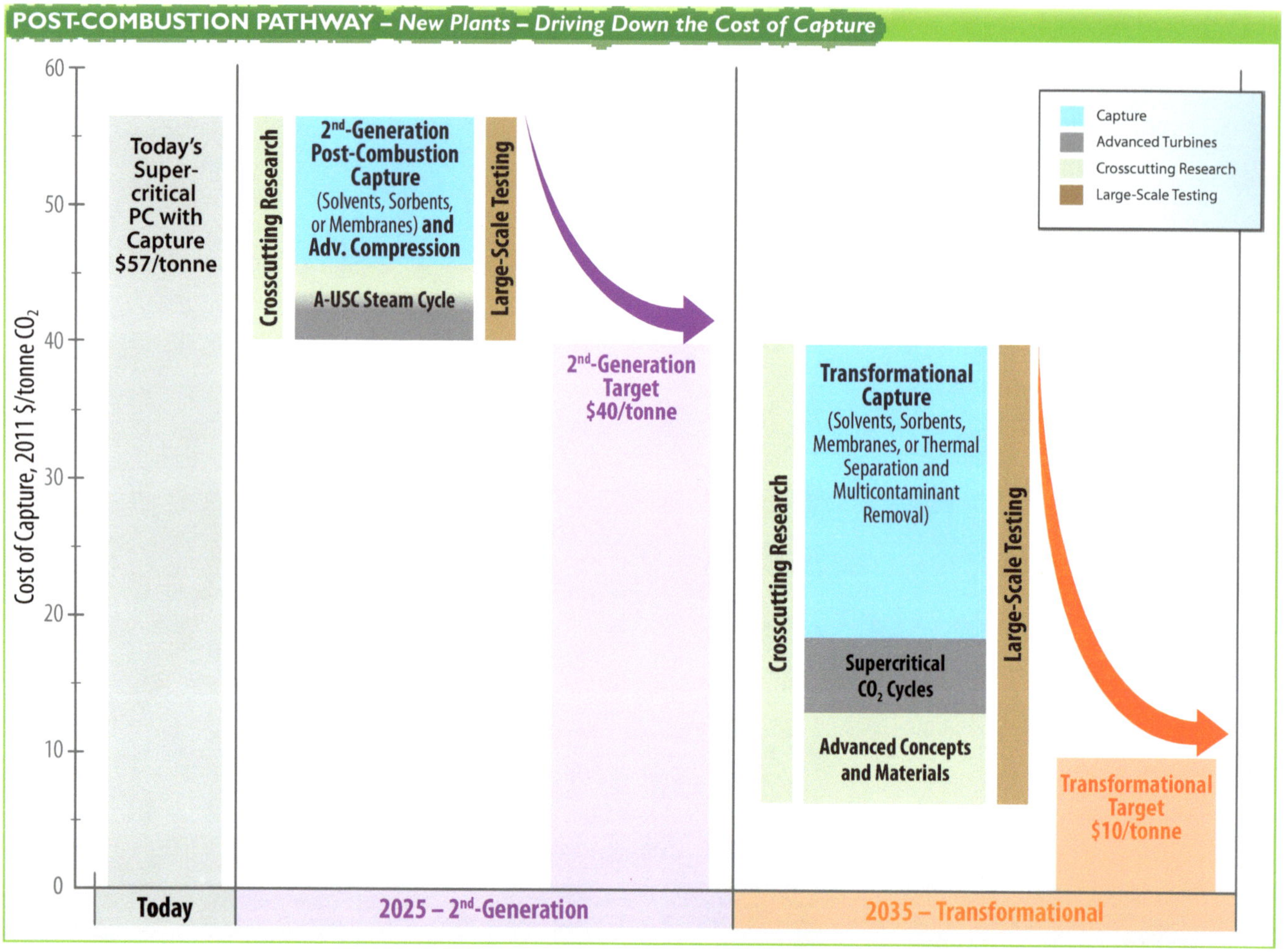

Figure 3-2. Targets for Technology Contributions to Overall CCRP Cost of Capture Goals—Post-Combustion New Plants Pathway

The applicability of post-combustion capture technologies to existing plants differentiates it from other technologies designed to reduce carbon emissions. In 2010, coal-based power plants generated approximately 47 percent of the electricity in the United States, and coal is expected to continue to play a critical role in powering the nation's electricity generation for the foreseeable future. DOE's Energy Information Administration (EIA) projects that by 2035 over 270 GW of total electricity generation capacity will be coal-based and made up largely of plants currently in existence (EIA, 2012).

All segments of U.S. society rely heavily on America's existing multibillion-dollar investment in its highly reliable and affordable coal-based energy infrastructure. Energy production infrastructure is expensive to create and takes time to put in place in proper balance with market demand. Global competitiveness requires that investments in such infrastructure be prudently made and placed into reliable, long-term, high-capacity service. Making the "right" investments, both in terms of type and amount, keeps America competitive today and has significant upside ramifications to the long-term health of the U.S. economy. Making improvements in the use of the nation's current fossil energy production, conversion, and distribution infrastructure is one of the most immediate, direct, and effective means of supporting the nation's economy, environment, and energy-security needs.

Given the importance of the existing fleet of coal-based power production, the largest market for post-combustion capture technology is in power plant retrofits. Analyses have been conducted to establish appropriate goals for retrofit applications. In the case of 2nd-Generation technologies, CO_2 emissions for existing plants are assumed to remain unregulated and decisions to retrofit are based purely on the cost-effectiveness of generating revenue through CO_2 use, i.e., CO_2-EOR. In the case of Transformational technologies, it is assumed that existing plants face a "retire or retrofit" decision under a carbon emission regulation scenario that includes a price on CO_2 emissions. Given these assumptions, CCS retrofits can compete with new baseload power generation at $45/tonne CO_2 captured for 2nd-Generation technologies and at $30/tonne CO_2 captured for transformational technologies (Figure 3-3). Their competitive advantage derives from the fact that they require only the incremental costs associated with CO_2 capture. The initial capital investment for a CCS retrofit is 25–30 percent of a new plant because the balance of the plant already exists and is paid off.

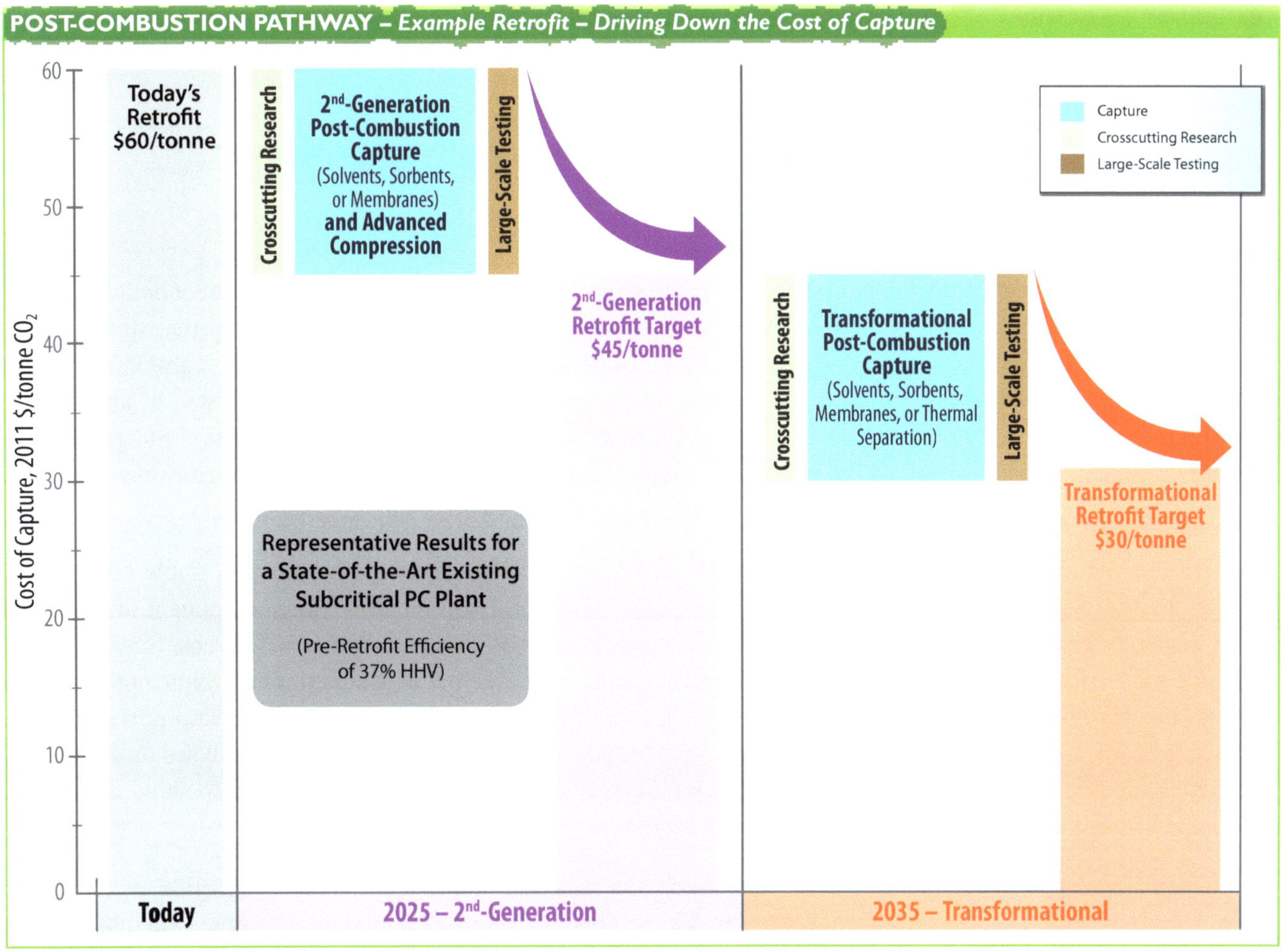

Figure 3-3. Goals for Post-Combustion Retrofits

The Carbon Capture program goals are summarized in Table 3-2 for post-combustion and pre-combustion technologies and are stated in terms of their contribution to a reduction in cost of CO_2 capture measured as $/tonne. For example, the goal for 2nd-Generation post-combustion capture technologies is a $12/tonne contribution to the reduction in cost of CO_2 capture for a new plant. The contribution to reduction in cost of CO_2 capture is measured relative to today's 1st-Generation technologies that have a cost of approximately $57/tonne of CO_2 captured. As a result, a 2nd-Generation post-combustion technology that meets the $12/tonne contribution to reduction in cost of CO_2 capture goal, when combined with an advanced ultra-supercritical boiler that provides a $5/tonne contribution, would result in a plant with an overall total cost of $40/tonne of CO_2 captured.

Table 3-2. **Carbon Capture Program Goals**

Technology Area	Contribution to Reduction in Cost of CO_2 Capture*	
	2nd-Generation Technology	*Transformational Technology*†
Post-Combustion Capture (new plant)	$12/tonne	$22/tonne
Post-Combustion Capture (existing plant retrofit)	$15/tonne	$15/tonne
Pre-Combustion Capture (new plant)	$5/tonne	$5/tonne

NOTES:

* Cost of CO_2 capture relative to advanced ultra-supercritical power plant without CCS for new plant and subcritical power plant without CCS for existing plant retrofit. All costs are in 2011 dollars. Includes approximately $1/tonne contribution from advanced compression technology.

† Transformational technology contributions shown are in addition to those derived from 2nd-Generation technologies. Thus the total cost reductions for Transformational technologies would be $34/tonne for post-combustion capture new plants, $30/tonne for post-combustion capture existing plant retrofits, and $10/tonne for pre-combustion capture.

3.2 BENEFITS

Coal-fired electricity generation with carbon capture has the potential to provide significant economic, environmental, and technical benefits. Coal is an abundant domestic fuel source with a stable price history that supports the U.S. economy, a resurgent industry, and even U.S. exports. Coal-based power-generation systems integrated with advanced technologies to improve process efficiency and reduce costs are being developed by DOE and will be able to generate power with greater than 90 percent carbon capture. Carbon captured from advanced combustion plants can be compressed, transported via pipeline, and injected into a depleted oil reservoir for EOR, thereby increasing the production of domestic oil. Alternatively, captured carbon can be used as feedstock for value-added products.

Furthermore, natural gas prices are currently low. However, historically natural gas has not had stable prices, and most predictions for natural gas prices have not been accurate. Coal is, and will remain, a key component in the U.S. electricity-generating portfolio. For the economy to be strong there must be enough continuous low-cost fossil-based power available for the foreseeable future. Advanced electricity-generation systems with superior environmental performance, through the development of advanced, highly efficient, low-cost technologies to convert coal into power with carbon capture, can fill this role. Industry and DOE have performed numerous techno-economic analyses on how advanced coal-based electricity-generation systems compete with other technologies. These analyses also show the advantages anticipated from the ongoing DOE-supported R&D program.

The development and deployment of new technologies for near-zero-emissions power production will result in the United States becoming a key leader in these technologies. This will create new, high paying domestic jobs to manufacture and oversee the deployment and operation of these next-generation advanced combustion plants. The capability to produce low-cost, coal-based electricity while eliminating nearly all air pollutants and potential GHG emissions makes carbon capture one of the most promising technologies for energy plants of the future.

THIS PAGE INTENTIONALLY LEFT BLANK

CHAPTER 4: TECHNICAL PLAN

4.1 INTRODUCTION

The Carbon Capture program consists of two core research Technology Areas: (1) Post-Combustion Capture and (2) Pre-Combustion Capture. Current R&D efforts conducted within the Carbon Capture program are focused on development of advanced solvents, sorbents, and membranes for both the Post-Combustion and Pre-Combustion Technology Areas. The subsections below describe the general characteristics of each key technology and provide details regarding different research focus areas associated with the key technologies.

4.2 POST-COMBUSTION CAPTURE

The Post-Combustion Technology Area includes three key technologies:

- Solvents
- Sorbents
- Membranes

The technical characteristics of each of these technologies are presented below, along with the R&D approach for each technology and associated performance targets and measures. In addition, a technology development timeline has been prepared.

4.2.1 BACKGROUND

Post-combustion CO_2 capture refers to removal of CO_2 from the flue gas produced from fossil fuel combustion. Although primarily applicable to conventional coal-, oil-, or gas-fired power plants, this approach could also be applicable to IGCC and natural gas combined cycle flue gas capture.

A simplified process schematic of post-combustion CO_2 capture is shown in Figure 4-1. In a coal-fired power plant, fuel is burned with air in a boiler to produce steam that drives a turbine/generator to produce electricity. Flue gas from the boiler consists mostly of N_2 and CO_2. The CO_2 capture process would be located downstream of the plant's conventional pollutant controls. The current 1st-Generation CO_2 capture technology—chemical absorption with aqueous monoethanolamine—is capable of achieving 90 percent CO_2 capture. However, it requires the extraction of a relatively large volume of low-pressure steam from the power plant's steam cycle, which decreases the gross electrical generation of the plant. The steam is required for release of the captured CO_2 and regeneration of the solvent. Additionally, separating CO_2 from this flue gas is challenging for several reasons: a high volume of gas must be treated (≈2 million cubic feet per minute for a 550-MWe plant), the CO_2 is dilute (between 12 and 14 percent CO_2), the flue gas is at atmospheric pressure, trace impurities (PM, SO_2, NO_x, etc.) can degrade chemical scrubbing agents, and compressing captured CO_2 from near-atmospheric pressure to pipeline pressure (about 2,200 psia) requires a large auxiliary power load.

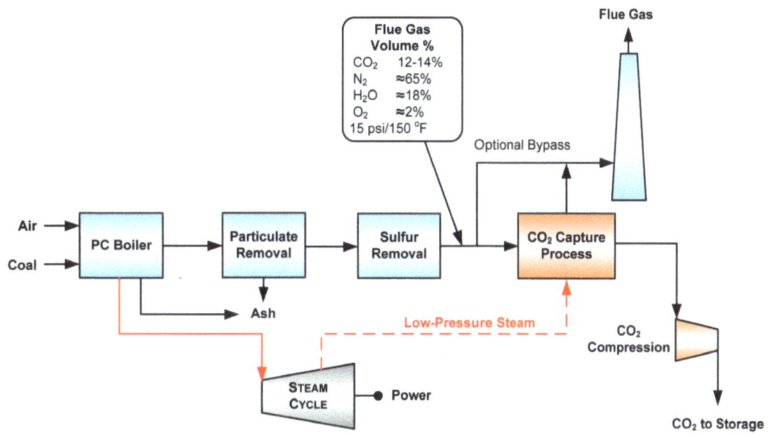

Figure 4-1. Process Schematic of Post-Combustion Capture

4.2.2 TECHNICAL DISCUSSION

DOE/NETL is currently funding the development of advanced post-combustion CO_2 capture technologies that have the potential to provide step-change improvements in both cost and performance as compared to the current state-of-the-art solvent-based processes. The R&D effort for post-combustion applications is focused on advanced solvents, solid sorbents, and membrane-based systems. In addition, hybrid technologies that combine attributes from multiple key technologies (e.g., solid sorbent material embedded onto a membrane-style contactor) are being investigated. This section describes the research focus for each of the three Post-Combustion Capture key technologies (Figure 4-2).

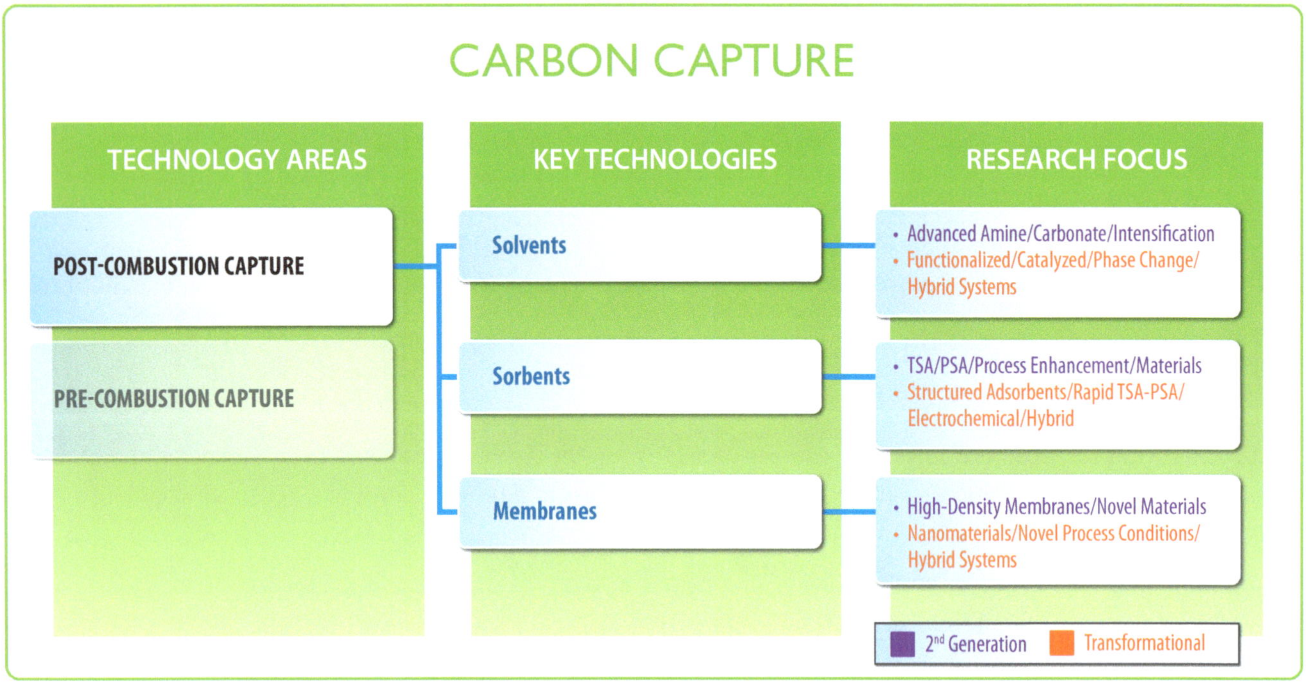

Figure 4-2. Key Technologies and Associated Research Focus for Post-Combustion Capture

SOLVENTS

Solvent-based CO_2 capture involves chemical or physical absorption of CO_2 from flue gas into a liquid carrier. The absorption liquid is regenerated by increasing its temperature or reducing its pressure. Research projects in this key technology focus on the development of low-cost, non-corrosive solvents that have a high CO_2 loading capacity, improved reaction kinetics, low regeneration energy, and resistance to degradation. In addition, considerable effort is being applied to development of process design and integration that leads to decreased capital and operating costs and enhanced performance. Transformational technologies that may be pursued include both switchable and non-switchable ionic liquids, catalyzed absorption that accelerates CO_2 uptake in solvents with lower regeneration energies, solvents that change phase in the presence of CO_2, and hybrid systems.

ION Engineering's Ionic Liquid Bench-Scale Test Equipment

As an example, the development of ionic liquids is being conducted by several organizations including the University of Notre Dame, General Electric, Battelle Pacific Northwest Division, and ION Engineering. Ionic liquids include a broad category of salts that can dissolve gaseous

CO_2 and are stable at temperatures up to several hundred degrees Centigrade. Since ionic liquids are physical solvents, less energy is required for regeneration compared to today's conventional chemical solvents. However, the ionic liquid working capacity still needs to be significantly improved to meet cost targets. One possible drawback is that the viscosities of many ionic liquids are relatively high upon CO_2 absorption compared to those associated with conventional solvents, perhaps adversely affecting the energy requirement to pump ionic liquids in a conventional adsorption/stripping process. Although the production cost for newly synthesized ionic liquids is high, the cost could be significantly lower when produced on a commercial scale.

SORBENTS

Solid sorbents—including sodium and potassium oxides, zeolites, carbonates, amine-enriched sorbents, and metal organic frameworks—are being explored for post-combustion CO_2 capture. A temperature or pressure swing facilitates sorbent regeneration following chemical and/or physical adsorption. However, a key attribute of CO_2 sorbents (compared with solvent-based systems) is that no water is present, thereby reducing sensible heating and stripping energy requirements. Possible configurations for contacting the flue gas with the sorbents include fixed, moving, and fluidized beds. Research projects in this key technology focus on developing sorbents with the following characteristics: low-cost raw materials, thermal and chemical stability, low attrition rates, low heat capacity, high CO_2 adsorption capacity, and high CO_2 selectivity. Another important focus of this research is to develop cost-effective process equipment designs that are tailored to the sorbent characteristics. Transformational concepts being considered include structured solid adsorbents (e.g., metal organic frameworks), enhanced pressure swing adsorption (PSA) and temperature swing adsorption (TSA) processes, hybrid systems, and electrochemical technologies.

ADA-ES 1-kW Pilot-Scale Test Equipment

In one project, in 2010–2011 ADA-Environmental Solutions (ADA-ES) conducted slipstream field testing of four supported amine sorbents, which had performed well during laboratory-scale screening. Two of these sorbents were developed and patented by NETL in-house researchers. ADA-ES was then awarded a follow-up project to design and construct a 1-MW pilot-scale plant to demonstrate solid sorbent-based post-combustion CO_2 capture technology to reduce uncertainty of scaleup and accelerate the path to commercialization. The field testing is scheduled to be conducted in 2014 at Southern Company's Plant Miller. Results of the pilot-scale testing will be used to prepare detailed designs and cost estimates for industrial- and utility-scale CO_2 capture applications.

DOE/NETL is also sponsoring the development of unique hybrid approaches for using sorbents. In one project, Georgia Tech Research Corporation is developing a process based on rapid TSA using polymer/supported amine composite hollow fibers. Another project is being conducted by the University of North Dakota to develop a process known as "Capture from Existing Coal-Fired Plants by Hybrid Sorption Using Solid Sorbents Capture" (CACHYS™). The technology uses a regenerable metal carbonate-based sorbent with a high CO_2 loading capacity coupled with a process design that results in a low regeneration energy penalty.

MEMBRANES

Membrane-based CO_2 capture uses permeable or semi-permeable materials that allow for the selective transport and separation of CO_2 from flue gas. Generally, gas separation is accomplished by some physical or chemical interaction between the membrane and the gas being separated, causing one component in the gas to permeate through the membrane faster than another. Usually the selectivity of the membrane is insufficient to achieve the desired purities and recoveries. Therefore multiple stages and recycle streams may be required in an actual operation, leading to increased complexity, energy consumption, and capital costs. Also under development are gas absorption membrane technologies where the separation is caused by the presence of an absorption liquid on one side of the membrane that selectively removes CO_2 from a gas stream on the other side of the membrane. Research projects in this key technology address technical challenges to the use of membrane-based systems, such as large flue gas volume, relatively low CO_2 concentration, low flue gas pressure, flue gas contaminants, and the need for high membrane surface area. The Department's research focus for post-combustion membranes includes development of low-cost, durable membranes that have improved permeability and selectivity, thermal and physical stability, and tolerance to contaminants in combustion flue gas. Transformational membrane technologies under investigation include hybrid systems, novel process conditions (e.g., systems that operate at sub-ambient temperatures), and nanomaterials.

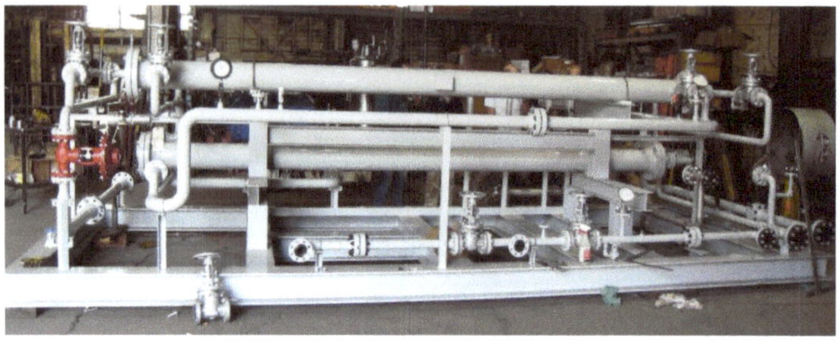

MTR's 0.05 MW Membrane Test Skid

MTR has been involved in a series of projects to develop a polymeric membrane and associated process for post-combustion CO_2 capture applications. MTR has made significant advances in membrane development, process configuration, and scaleup. The use of commercially available membranes for post-combustion CO_2 capture was previously considered impractical due to the large membrane area required for separation because of the low partial pressure of CO_2 in flue gas. However, MTR is using a twofold approach to address this issue: (1) the development of high-permeance membranes to reduce the required membrane area and associated capital cost and (2) the use of incoming combustion air in a countercurrent/sweep module design to generate separation driving force and reduce the need for vacuum pumps and the associated parasitic energy cost. In addition to improving membrane performance, improved membrane manufacturing techniques and materials have been developed. The current project includes conducting slipstream laboratory-scale (0.05 MWe) and small pilot-scale (1 MWe) tests using full-scale commercial membrane modules.

SUMMARY OF TECHNOLOGY ADVANTAGES AND CHALLENGES

The advantages and challenges associated with post-combustion capture using solvents, sorbents, and membranes are presented in Table 4-1. The approach developed for addressing those challenges is described in the following sections.

Table 4-1. **Post-Combustion Technology Advantages and Challenges**

CO_2 Capture Technology	Advantages	Challenges
Solvents	• Chemical solvents provide a high chemical potential (or driving force) necessary for selective capture from streams with low CO_2 partial pressure • Wet-scrubbing allows good heat integration and ease of heat management (useful for exothermic absorption reactions)	• Tradeoff between heat of reaction and kinetics; current solvents require a significant amount of steam to reverse chemical reactions and regenerate the solvent, which derates power plant • Energy required to heat, cool, and pump non-reactive carrier liquid (usually water) is often significant • Vacuum stripping can reduce regeneration steam requirements, but is expensive

Table 4-1. **Post-Combustion Technology Advantages and Challenges**

CO_2 Capture Technology	Advantages	Challenges
Solid Sorbents	• Chemical sites provide large capacities and fast kinetics, enabling capture from streams with low CO_2 partial pressure • Higher capacities on a per mass or volume basis than similar wet-scrubbing chemicals • Lower heating requirements than wet-scrubbing in many cases (CO_2 and heat capacity dependent) • Dry process—less sensible heating requirement than wet-scrubbing process	• Heat required to reverse chemical reaction (although generally less than in wet-scrubbing cases) • Heat management in solid systems is difficult, which can limit capacity and/or create operational issues when absorption reaction is exothermic • Pressure drop can be large in flue gas applications • Sorbent attrition
Membranes	• No steam load • No chemicals • Simple and modular designs • "Unit operation" versus complex "process"	• Membranes tend to be more suitable for high-pressure processes such as IGCC • Tradeoff between recovery rate and product purity (difficult to attain both high recovery rate and high purity) • Requires high selectivity (due to CO_2 concentration and low pressure ratio) • Poor economy of scale • Multiple stages and recycle streams may be required

4.2.3 TECHNOLOGY TIMELINE

The Carbon Capture program has adopted an aggressive timeline for developing 2nd-Generation and Transformational post-combustion capture technologies. Figure 4-3 presents an overview of the timeline for technology development. The 2nd-Generation technology timelines are shown in shades of purple and the Transformational technology timelines are shown in shades of orange. Each timeline in the figure consists of three major RD&D phases: (1) research testing, (2) pilot-scale testing, and (3) demonstration-scale testing. The research phase includes laboratory- and bench-scale testing, while the pilot-scale testing includes small pilot-scale testing (0.5–5 MW equivalent) and large pilot-scale testing (10–25 MW equivalent). The post-combustion CO_2 capture technologies that are successfully tested at the laboratory/bench-scale will advance to pilot-scale slipstream testing using actual flue gas at host coal-fired power plants or large test facilities such as DOE/NETL's NCCC. The laboratory-scale through large pilot-scale testing is planned to be conducted through DOE/NETL funding, while it is anticipated that the demonstration-scale testing will be conducted through private industry funding. As a result of these efforts, 2nd-Generation and Transformational post-combustion capture technologies will be ready for demonstration-scale testing after 2020 and 2030, respectively.

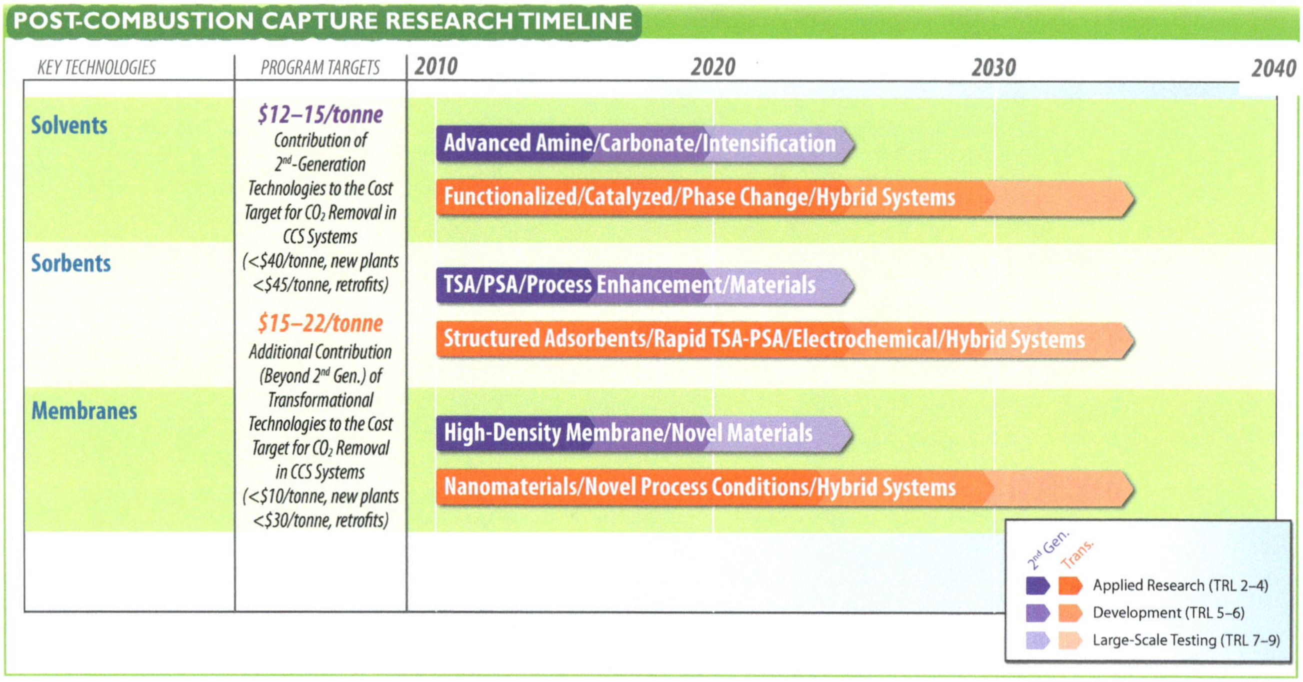

Figure 4-3. Post-Combustion Capture Development Timeline

4.3 PRE-COMBUSTION CAPTURE

The Pre-Combustion Technology Area includes three key technologies:

- Solvents
- Sorbents
- Membranes

The technical characteristics of each of these technologies are presented below along with the R&D approach for each technology and associated performance targets and measures. In addition, a technology development timeline has been prepared.

4.3.1 BACKGROUND

Pre-combustion capture is mainly applicable to IGCC power plants, where fuel is converted into gaseous components by applying heat under pressure in the presence of steam and substoichiometric O_2. A simplified process schematic for pre-combustion CO_2 capture is shown in Figure 4-4.

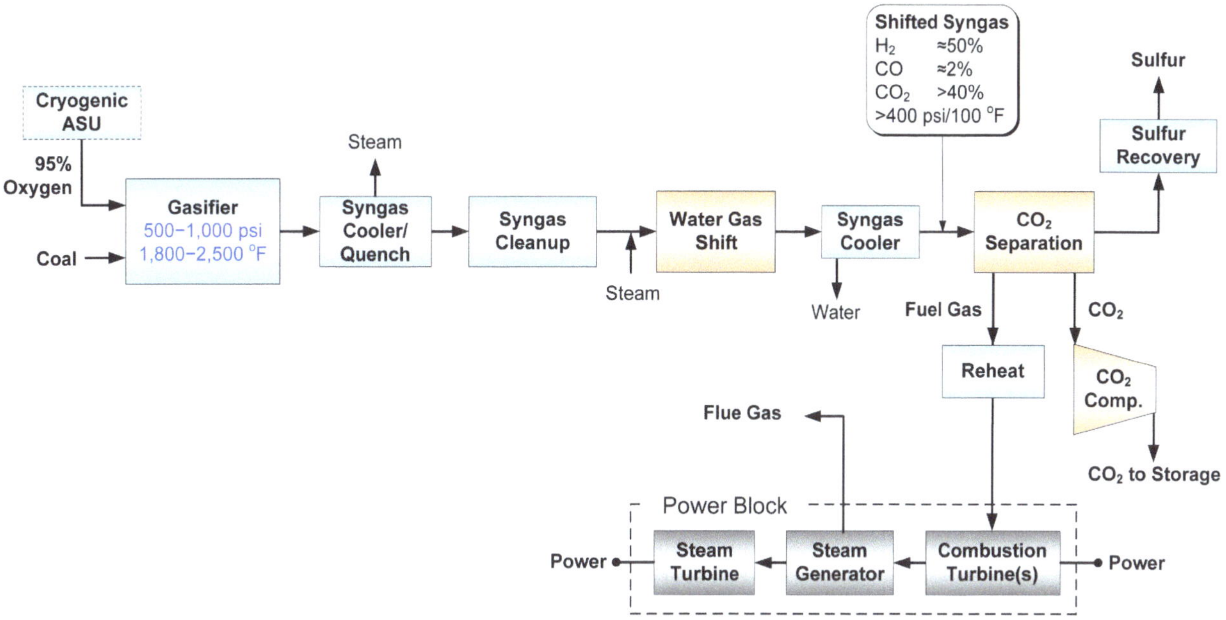

Figure 4-4. Process Schematic of Pre-Combustion Capture

By carefully controlling the amount of O_2, only a portion of the fuel burns to provide the heat necessary to decompose the remaining fuel and produce syngas, a mixture of H_2 and carbon monoxide, along with minor amounts of other gaseous constituents (e.g., sulfur). To enable pre-combustion capture, the syngas is further processed in a WGS reactor, which converts carbon monoxide into CO_2 while producing additional H_2, thus increasing the CO_2 and H_2 concentrations. An acid-gas removal system can then be used to separate CO_2 from the H_2. After WGS, the CO_2 in syngas is present at relatively higher concentrations than in flue gas. Also, the syngas is at higher pressure relative to flue gas. These characteristics make pre-combustion carbon capture relatively simpler and less expensive compared to post-combustion carbon capture. After CO_2 removal, the H_2 is used as a fuel in a combustion turbine combined cycle to generate electricity or other useful, high-value products.

The current state-of-the-art pre-combustion CO_2 capture technologies that could be applied to IGCC systems—the glycol-based Selexol™ process and the methanol-based Rectisol® process—employ physical solvents that preferentially absorb CO_2 from the syngas mixture. Today, these technologies are not considered cost-effective for application to IGCC power plants.

4.3.2 TECHNICAL DISCUSSION

DOE/NETL is currently funding the development of advanced pre-combustion CO_2 capture technologies that have the potential to provide step-change improvements in both cost and performance as compared to the current state-of-the-art solvent-based processes. The R&D effort for pre-combustion applications is focused on advanced solvents, solid sorbents, and membrane-based systems for the separation of H_2 and CO_2. In addition, hybrid technologies that combine attributes from multiple technologies (e.g., CO_2 separation and WGS) are being investigated. This section describes the research focus for each of the three Pre-Combustion Capture key technologies (Figure 4-5).

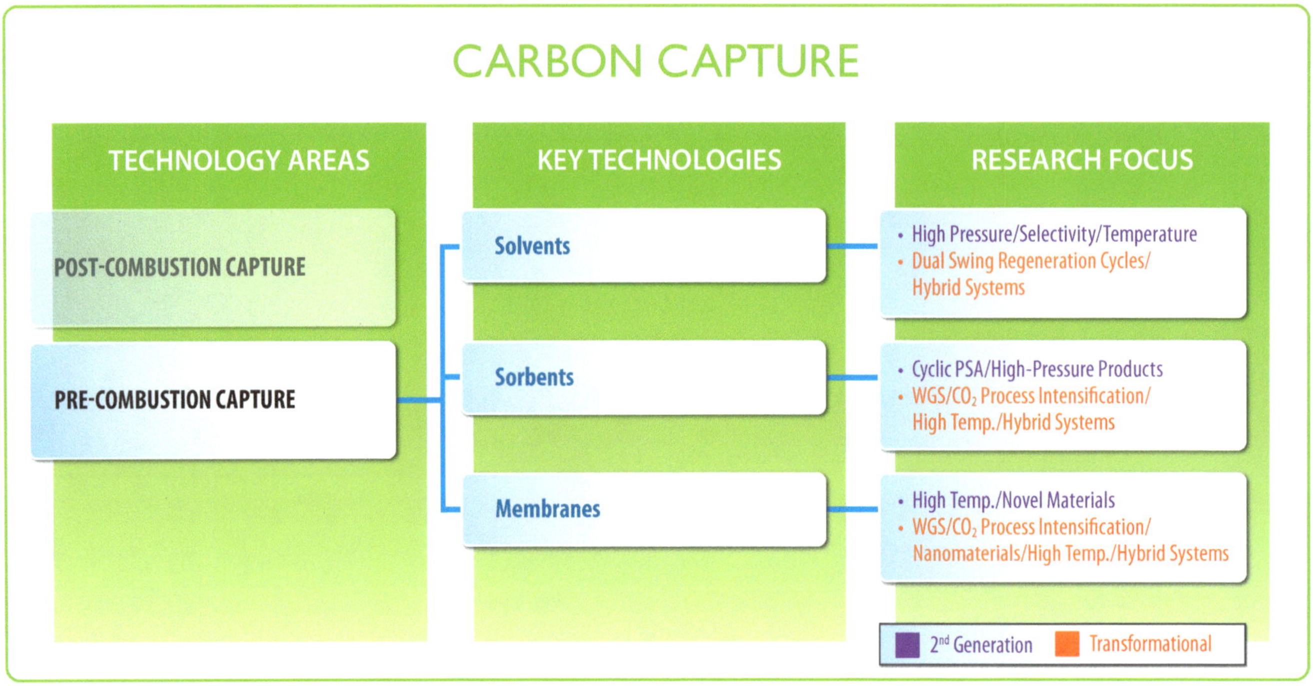

Figure 4-5. Key Technologies and Associated Research Focus for Pre-Combustion Capture

SOLVENTS

Solvent-based CO_2 capture involves chemical or physical absorption of CO_2 from flue gas into a liquid carrier. As the name implies, a chemical solvent relies on a chemical reaction for absorption, whereas a physical solvent selectively absorbs CO_2 without a chemical reaction. The main benefit of a physical solvent, as compared to a chemical solvent, is that it requires less energy for regeneration. However, chemical solvents offer the advantages of increased mass transfer driving force into solution, increased acid gas selectivity, and the potential to generate the CO_2 at elevated pressure. Challenges associated with solvent-based pre-combustion CO_2 capture include modifying regeneration conditions to recover the CO_2 at a higher pressure, improving selectivity to reduce H_2 losses, and developing a solvent that has a high CO_2 loading at a higher temperature to improve IGCC efficiency. Transformational technologies being considered include combining temperature-swing and pressure-swing regeneration to lower cost and energy penalty and development of hybrid systems.

Pre-combustion solvent R&D activities focus on a number of research objectives that address solvent technology challenges, including increasing CO_2 loading capacity and reaction kinetics coupled with decreasing regeneration energy. DOE/NETL's Office of Research and Development (ORD) is evaluating the use of ionic liquids as physical solvents for CO_2 capture in IGCC applications. Ionic liquids can absorb CO_2 at elevated temperature, providing a potential option to combine CO_2 capture with warm syngas cleanup.

SORBENTS

The materials, regeneration characteristics, and process configurations for pre-combustion sorbents are similar to those described for post-combustion sorbents but applied to the unique conditions of IGCC systems. Research projects in pre-combustion sorbent technology focus on the development of sorbents with the following characteristics: high adsorption capacity, resistance to attrition over multiple regeneration cycles, and good CO_2 separation and selectivity performance at the high temperatures encountered in IGCC systems to avoid the need for syngas cooling. Another important focus of the research is to develop cost-effective process equipment designs that are tailored to the sorbent characteristics. Some of the current pre-combustion sorbents under development include activated

TDA's Carbon-Based Sorbent Slipstream Test Skid at NCCC

carbon, alumina, calcium carbonate, and magnesium oxide. Transformational technologies include integrating capture directly with the WGS reaction to help drive equilibrium toward CO_2 and H_2 production while eliminating the need for syngas cooling and development of hybrid systems.

The advantage of an adsorption process is that some solid sorbents can be used at a high temperature. In a pre-combustion application, this is important since high-temperature (above 500 °F) CO_2 capture combined with warm/humid gas sulfur cleanup would eliminate syngas reheating and thus improve the overall thermal efficiency of the IGCC power plant.

In one project, TDA is developing a proprietary sorbent that consists of a carbon support modified with surface functional groups that remove CO_2 via strong physical adsorption. The CO_2 surface interaction is strong enough to allow operation at elevated temperatures. The process uses two (or more) beds that switch positions between adsorption and regeneration. In addition to the conventional pressure and temperature swing operation, the sorbent can be regenerated under near isothermal and isobaric conditions, while the driving force for separation is provided by a swing in CO_2 concentration. Slipstream field tests have been conducted at ConocoPhillips' Wabash River IGCC Plant.

MEMBRANES

Los Alamos National Laboratory's PBI Membrane

As with sorbents, the general characteristics of pre-combustion membranes are similar to those for post-combustion. Research is being conducted on CO_2 selectivity and permeability in pre-combustion systems, thermal and hydrothermal stabilities of the membrane, as well as other physical and chemical properties. Scaleup studies must determine the potential for lower cost and efficient operation in integrated systems. Large-scale manufacturing methods for defect-free membranes and modules must be developed. Better methods are needed to make high-temperature, high-pressure seals. Several advanced membrane technology options are under development. Membrane designs include metallic, polymeric, or ceramic materials operating at elevated temperatures, with a variety of chemical and/or physical mechanisms that provide separation. Transformational technologies include integration of a membrane-based system with WGS, high density and pressure nanoscale membranes, high-temperature/high-pressure seals, process intensification, and hybrid systems.

SRI International is testing a high-temperature polybenzimidazole (PBI) polymer membrane developed by Los Alamos National Laboratory. The membrane consists of hollow-fiber PBI, which is chemically and thermally stable at temperatures up to 450 °C and pressures up to 55 atm (800 pounds per square inch gauge). This characteristic permits the use of a membrane for CO_2 capture without requiring additional syngas cooling after the WGS reactor, which would increase plant efficiency. In addition, the CO_2 is recovered at high pressure, decreasing CO_2 compression requirements.

In another project, MTR is developing a new polymer membrane and membrane separation process. The first phase of the project focused on materials development with an emphasis on identifying high-temperature polymers with desirable H_2/CO_2 separation properties. Promising polymers were then fabricated into industrial flat-sheet composite membranes and bench-scale tested with simulated syngas mixtures. In the second phase of the project, optimized membrane materials were down-selected and fabricated into composite membranes, which were formed into laboratory-scale, spiral-wound modules and tested with simulated syngas. The third phase of the project included field testing the membranes at the National Carbon Capture Center using actual syngas. Based on this successful testing, additional scaleup testing for a commercial-scale membrane module was conducted in 2011.

SUMMARY OF TECHNOLOGY ADVANTAGES AND CHALLENGES

The advantages and challenges associated with pre-combustion capture using solvents, sorbents, and membranes are presented in Table 4-2. The approach developed for addressing these challenges is described in the following sections.

Table 4-2. **Pre-Combustion Technology Advantages and Challenges**

CO_2 Capture Technology	Advantages	Challenges
Physical Solvents	• CO_2 recovery does not require heat to reverse a chemical reaction • Common for same solvent to have high H_2S solubility, allowing for combined CO_2/H_2S removal • System concepts in which CO_2 is recovered with some steam stripping, rather than flashed, and delivered at a higher pressure may optimize processes for power systems	• CO_2 pressure is lost during flash recovery • Must cool down syngas for CO_2 capture, then heat it back up again and rehumidify for firing to turbine • Low solubilities can require circulating large volumes of solvent, resulting in large pump loads • Some H_2 may be lost with the CO_2
Solid Sorbents	• CO_2 recovery does not require heat to reverse a reaction • Common for H_2S to also have high solubility in the same sorbent, meaning CO_2 and H_2S capture can be combined • System concepts in which CO_2 is recovered with some steam stripping, rather than flashed, and delivered at a higher pressure may optimize processes for power systems	• CO_2 pressure is lost during flash recovery • Must cool syngas for CO_2 capture, then heat it back up again and rehumidify for firing to turbine • Some H_2 may be lost with the CO_2
H_2/CO_2 Membranes	H_2 or CO_2 Permeable Membrane • No steam load or chemical attrition H_2 Permeable Membrane Only • Can deliver CO_2 at high-pressure, greatly reducing compression costs • H_2 permeation can drive the carbon monoxide shift reaction toward completion—potentially achieving the shift at lower cost/higher temperatures	• Membrane separation of H_2 and CO_2 is more challenging than the difference in molecular weights implies • Due to decreasing partial pressure differentials, some H_2 will be lost with the CO_2 • In H_2 selective membranes, H_2 compression is required and offsets the gains of delivering CO_2 at pressure; in CO_2 selective membranes, CO_2 is generated at low pressure, requiring compression
Water-Gas-Shift Membranes	• Promote higher conversion of carbon monoxide and H_2O to CO_2 and H_2 than is achieved in a conventional WGS reactor • Reduce CO_2 capture costs • Reduce H_2 production costs • Increase net plant efficiency	• Single-stage WGS with membrane integration • Improved selectivity of H_2 or CO_2 • Optimize membranes for WGS reactor conditions

4.3.3 TECHNOLOGY TIMELINE

The Carbon Capture program has adopted an aggressive timeline for developing 2nd-Generation and Transformational pre-combustion capture technologies. Figure 4-6 presents an overview of the timeline for technology development based upon current program funding levels. The 2nd-Generation technology timelines are shown in shades of purple and the Transformational technology timelines are shown in shades of orange. Each timeline in the figure consists of three major RD&D phases: (1) research testing, (2) pilot-scale testing, and (3) demonstration-scale testing. The research phase includes laboratory- and bench-scale testing, while the pilot-scale testing includes small pilot-scale testing (0.1–0.5 MW equivalent) and large pilot-scale testing (1.5–6 MW equivalent). The pre-combustion CO_2 capture technologies that are successfully tested at the laboratory/bench-scale will advance to pilot-scale slipstream testing using actual syngas at host IGCC plants or large test facilities such as DOE/NETL's NCCC. The laboratory-scale through large pilot-scale testing is planned to be conducted through DOE/NETL funding, while it is anticipated that the demonstration-scale testing will be conducted through private industry funding. As a result of these efforts, 2nd-Generation and Transformational pre-combustion capture technologies will be ready for demonstration-scale testing after 2020 and 2030, respectively.

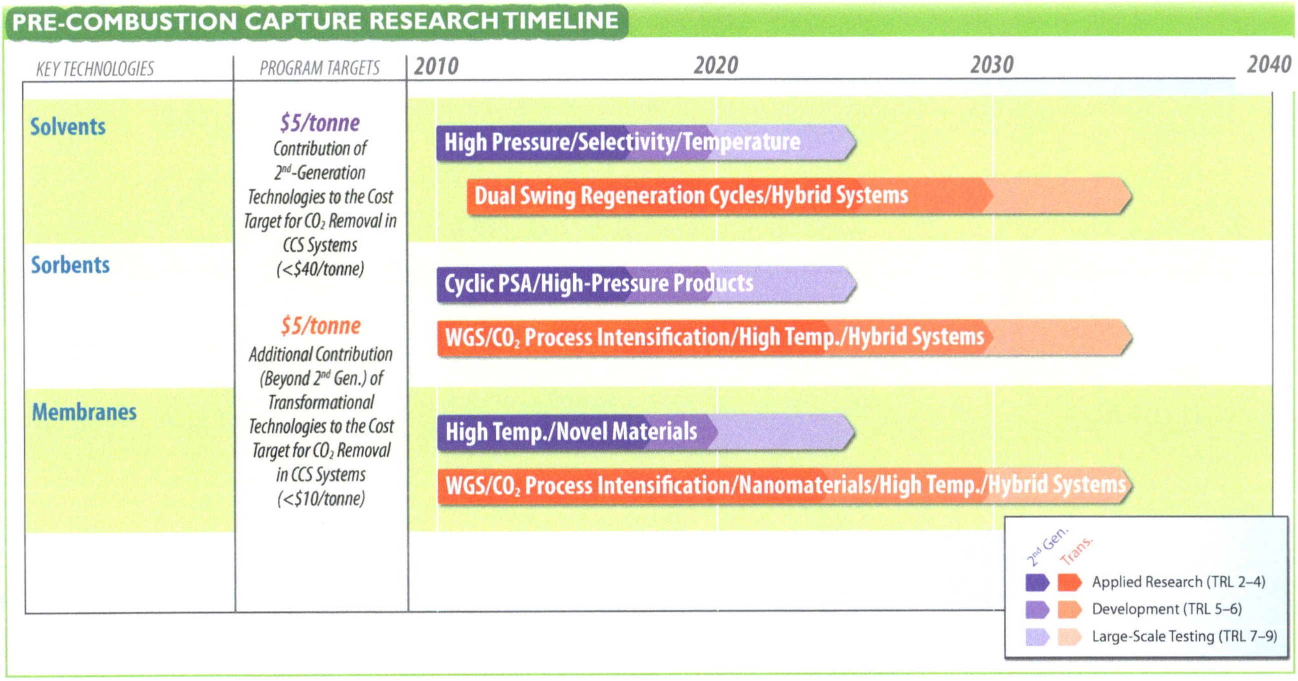

Figure 4-6. Pre-Combustion Capture Development Timeline

CHAPTER 5: IMPLEMENTATION AND COORDINATION PLAN

5.1 IMPLEMENTATION PLAN

Laboratory- and bench-scale testing of carbon capture systems is usually conducted with simulated flue or synthesis gas at relatively low gas flow rates. Upon completion of laboratory- and bench-scale testing, it is necessary to conduct pilot-scale slipstream testing using actual flue gas to determine potential adverse effects on the process from minor constituents in the coal that are present in the combustion flue gas or syngas. In addition, potential problems with excessive scaling, plugging, and/or corrosion of process equipment can only be evaluated and solutions developed via operating experience during long-term, pilot-scale slipstream testing. Slipstream testing will be conducted at host coal-fired PC or IGCC plants or at large test facilities such as DOE/NETL's National Carbon Capture Center.

The mission of the NCCC is to develop technologies that will lead to the commercialization of cost-effective, advanced coal-based power plants with CO_2 capture. The NCCC can test multiple projects in parallel with a wide range of test equipment sizes leading up to pre-commercial equipment sufficient to guide the design of full commercial-scale power plants. The NCCC is capable of testing both post- and pre-combustion technologies.

Southern Company's Plant Gaston power station provides the flue gas slipstream for the NCCC post-combustion CO_2 capture test facility. This flexible test module provides a site for testing technologies at a wide range of sizes and process conditions on coal-derived flue gas. The NCCC provides several parallel paths in order to test the candidate processes at the appropriate scale (Figure 5-1). For R&D projects that have been successfully tested at bench-scale in a research lab, the NCCC can provide a 1,000 lb/hr flue gas slipstream for screening tests. For technologies that have been successfully tested at the screening-scale, the NCCC provides a flue gas stream for pilot-scale testing. Two pilot test beds have been designed—a 5,000 lb/hr (0.5-MW equivalent) slipstream and a 10,000 lb/hr (1.0-MW equivalent) slipstream.

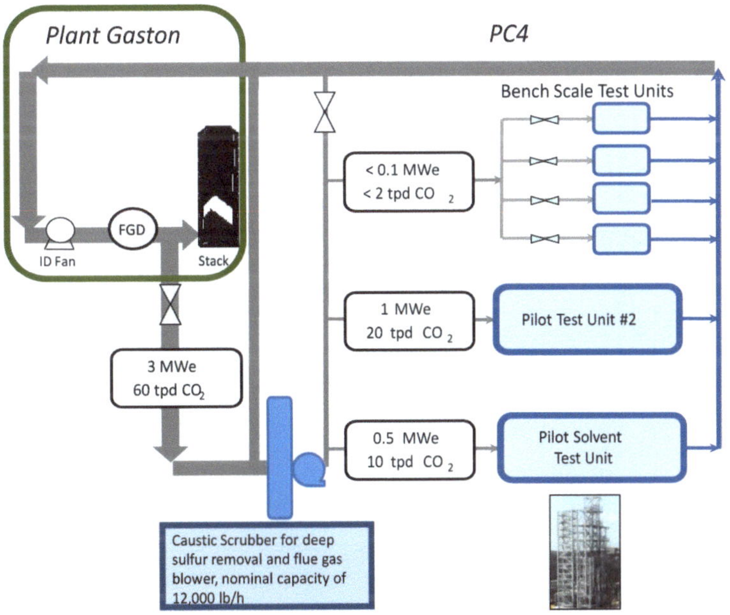

Figure 5-1. NCCC Post-Combustion Capture Test Facility

CHAPTER 5: IMPLEMENTATION AND COORDINATION PLAN

The backbone of the pre-combustion CO_2 capture technology development is a high-pressure flexible facility designed to test an array of solvents and contactors (Figure 5-2). Slipstreams are available with a range of gas flow rates and process conditions using coal-derived syngas for verification and scaleup of fundamental R&D capture projects.

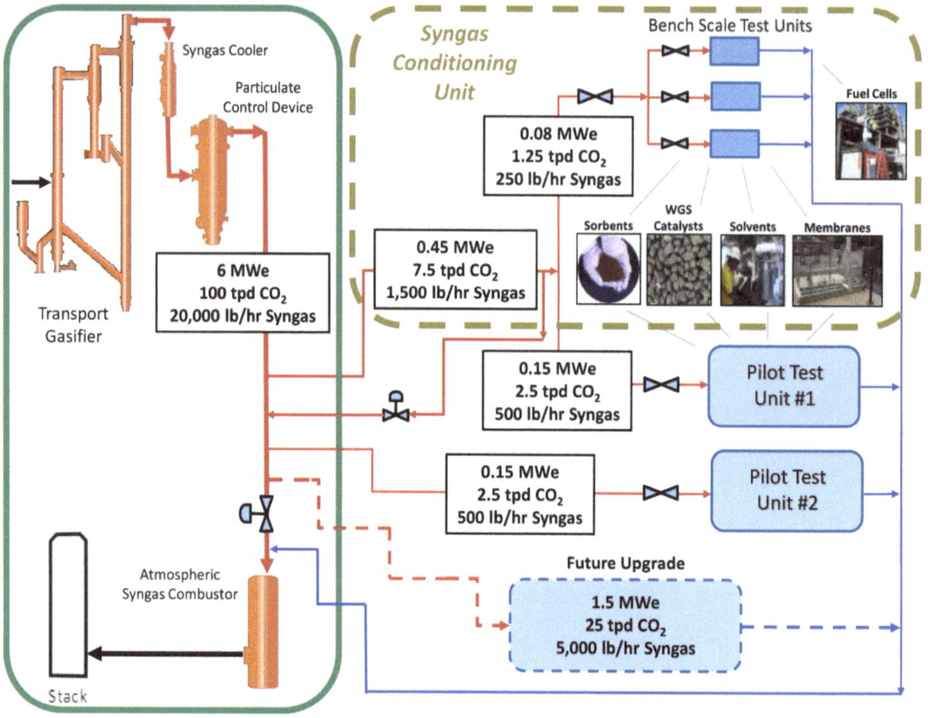

Figure 5-2. NCCC Pre-Combustion Capture Test Facility

5.2 CARBON CAPTURE PROGRAM ROADMAP

The Carbon Capture R&D program will be implemented as illustrated in the roadmap shown in Figure 5-3.

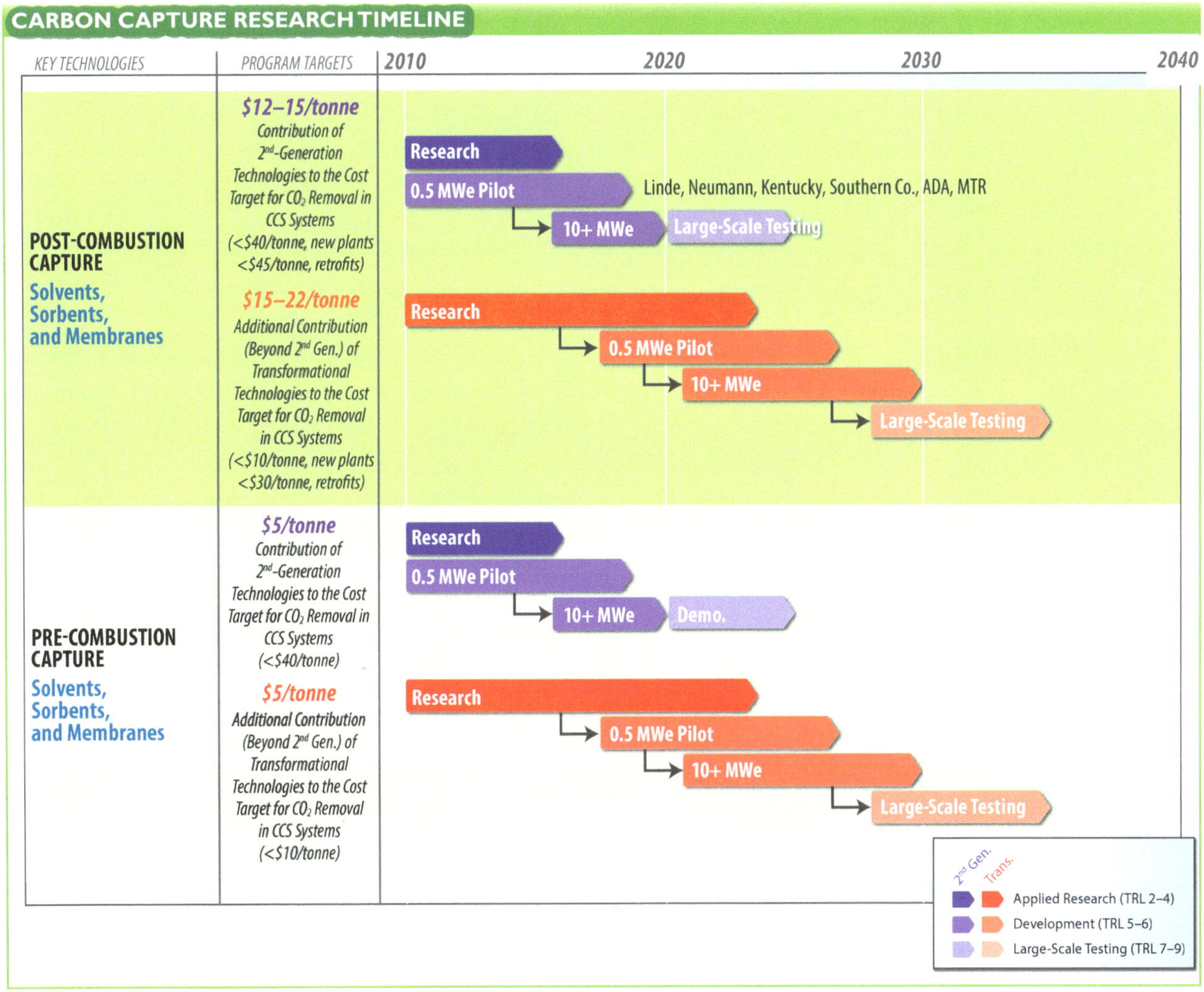

Figure 5-3. Carbon Capture Program RD&D Roadmap

For Post-Combustion Capture, laboratory- and bench-scale testing of 2nd-Generation technologies was initiated prior to 2010. Small pilot-scale field tests (0.5–5 MWe) were initiated for selected 2nd-Generation technologies in 2010. There are currently five projects being conducted at scales ranging from 0.5 to 1 MWe on three solvent-based technologies (Linde, Neumann Systems Group, and University of Kentucky), one sorbent-based technology (ADA-ES), and one membrane-based technology (MTR). In addition, heat integration testing is being conducted on a 25 MWe solvent-based capture technology (Southern Company). Small pilot-scale testing is anticipated to continue on these and other 2nd-Generation technologies through 2018, leading to large pilot-scale field testing (10–25 MW equivalent) of the most promising advanced CO_2 capture technologies beginning in 2015 and running through 2020. This schedule provides the opportunity to begin conducting demonstration-scale testing after 2020, and it is anticipated that some 2nd-Generation technologies would be ready for deployment after 2025.

For Pre-Combustion Capture, 2nd-Generation technology development efforts to date have focused on laboratory/bench-scale testing. One of the challenges associated with scaleup of Pre-Combustion Capture systems is that there will be only a handful of potential IGCC host sites within the United States operating on coal (including two currently in operation—Tampa Electric and Wabash) during this development period. Due to the small number of commercial installations, there are a limited number of test sites available to provide the synthesis gas necessary for conducting small and large pilot-scale testing. Commercial sites are also exhibiting reluctance to serve as host sites for pilot-scale testing because of concerns regarding potential adverse impacts on the facility's commercial IGCC operations. An alternative to limited commercial site availability is testing at the NCCC. However, this limits the testing that can be conducted to scales smaller than those that can be used for post-combustion systems. Initial small pilot-scale testing (0.1–0.5 MWe) is expected to be initiated near the end of 2013. Large pilot-scale testing (1.5–6 MWe) will follow over 2016–2020 at which point demonstration-scale testing is expected to be initiated. Deployment of 2nd-Generation technologies would begin sometime after 2025.

Laboratory- and bench-scale testing of Transformational post-combustion capture technologies was also initiated prior to 2010. The expectation is that the development of these Transformational technologies will follow the same general pattern as that for the 2nd-Generation technologies. However, since the Transformational technologies are at an earlier developmental stage, advancement of the technologies to small and large pilot scales will require additional time. Thus it is anticipated that small pilot-scale testing will be initiated in the 2014–2016 timeframe and continue through 2025. Large pilot-scale testing for the Transformational technologies farthest along the developmental path will likely be initiated in the 2020–2022 timeframe and will continue through 2030. Some technologies may be ready for demonstration-scale testing as early as the 2027 timeframe, and additional demonstrations are expected to continue through 2035. Deployment is expected to occur at some point after 2035. The development timeline for Transformational pre-combustion capture technologies will be similar to the timeline for post-combustion technologies. However the same limitations on scale will apply to these technologies as described above for the 2nd-Generation pre-combustion capture technologies.

In addition to the technologies already under development as part of the Carbon Capture program, it is anticipated that additional Transformational technologies will emerge from DOE's Advanced Research Projects Agency-Energy program (ARPA-E) that performs basic research on CO_2 capture technology under its Innovative Materials & Processes for Advanced Carbon Capture Technologies (IMPACCT) program. ARPA-E was organized in 2007 as the energy equivalent to the Department of Defense's Defense Advanced Research Projects Agency (DARPA). One of ARPA-E's objectives is to advance creative "out-of-the-box" Transformational energy research that industry by itself cannot or will not support due to its high risk, but where success would provide dramatic benefits for the nation. ARPA-E complements existing Carbon Capture program efforts by accelerating promising ideas from the basic research stage.

A program to aid in the technology development effort, the Carbon Capture Simulation Initiative (CCSI), was initiated in September 2010. The CCSI is designed to accelerate CO_2 capture technology development using advanced simulation and modeling techniques to develop lower cost, efficient industrial processes. By using simulations, DOE/NETL hopes to more quickly design appropriate processes for CO_2 capture and storage. This research is being shared among several national laboratories and universities—NETL and its Regional University Alliance (NETL-RUA) (comprising Carnegie Mellon University, Penn State University, University of Pittsburgh, Virginia Tech, and West Virginia University), Lawrence Berkeley National Laboratory, Los Alamos National Laboratory, Pacific Northwest National Laboratory, and Lawrence Livermore National Laboratory. Because of the early stage of this effort, the impact of this initiative on the CO_2 Capture R&D program timeline is unknown at this time.

The R&D required for the technology development effort will be facilitated through a series of competitive research solicitations. In 2013, a solicitation is planned calling for proposals to develop laboratory/bench and small pilot-scale technologies for both pre- and post-combustion capture. The laboratory/bench-scale projects will be for Transformational technologies whereas the small pilot-scale projects will be for 2nd-Generation technologies. Additional solicitations are anticipated through 2020, focusing on laboratory/bench-scale and small pilot-scale Transformational technologies as well as small and large pilot-scale 2nd-Generation technologies. Beyond that point, planning efforts have assumed that solicitations be released approximately every 2 years to support the development and scaleup of Transformational technologies. As the program is currently outlined, the result of the R&D effort for 2nd-Generation technologies will be five post-combustion and three pre-combustion capture systems ready for demonstration-scale testing in 2020. Similarly for Transformational technologies, by 2030 five post-combustion and three pre-combustion technologies will be ready for demonstration-scale testing.

The Carbon Capture program described above represents a comprehensive, multipronged technology R&D approach. R&D on a portfolio of technologies is being pursued to enhance the probability of success of research efforts that are operating at the boundaries of current scientific understanding. The research focus areas cover a wide scale, integrating advances and lessons learned from fundamental research, technology development, and demonstration-scale testing. The success of this effort will enable cost-effective implementation of carbon capture technologies.

5.3 COORDINATION WITH OTHER TECHNOLOGY AREAS

In support of the CCRP goals, the Carbon Capture, Advanced Energy Systems, and Crosscutting Research programs are conducting complementary R&D activities to develop technologies that enable low-cost electricity production while exceeding all environmental emission standards. Together, these technologies comprise a new generation of clean coal-based power systems capable of producing affordable electric power while capturing greater than 90 percent of the carbon normally emitted to the environment. Pre-Combustion Capture is highly dependent on the Gasification Systems Technology Area within the Advanced Energy Systems subprogram in order to accomplish CCRP goals. In turn, Gasification Systems is dependent on R&D being conducted as part of the Advanced Combustion Systems, Advanced Turbines, and Crosscutting Research programs. Integration of the different R&D activities facilitates the ability to meet overall goals.

APPENDIX A: DOE-FE TECHNOLOGY READINESS LEVELS

Table A-1. **Definitions of Technology Readiness Levels**

TRL	DOE-FE Definition	DOE-FE Description
1	Basic principles observed and reported	Lowest level of technology readiness. Scientific research begins to be translated into applied R&D. Examples might include paper studies of a technology's basic properties.
2	Technology concept and/or application formulated	Invention begins. Once basic principles are observed, practical applications can be invented. Applications are speculative, and there may be no proof or detailed analysis to support the assumptions. Examples are still limited to analytic studies.
3	Analytical and experimental critical function and/or characteristic proof of concept	Active R&D is initiated. This includes analytical studies and laboratory-scale studies to physically validate the analytical predictions of separate elements of the technology. Examples include components that are not yet integrated or representative. Components may be tested with simulants.
4	Component and/or system validation in laboratory environment	The basic technological components are integrated to establish that the pieces will work together. This is relatively "low fidelity" compared with the eventual system. Examples include integration of "ad hoc" hardware in a laboratory and testing with a range of simulants.
5	Laboratory scale, similar system validation in relevant environment	The basic technological components are integrated so that the system configuration is similar to (matches) the final application in almost all respects. Examples include testing a high-fidelity, laboratory-scale system in a simulated environment with a range of simulants.
6	Engineering/pilot scale, similar (prototypical) system demonstrated in a relevant environment	Engineering-scale models or prototypes are tested in a relevant environment. This represents a major step up from a TRL 5. Examples include testing an engineering scale prototype system with a range of simulants. TRL 6 begins true engineering development of the technology as an operational system.
7	System prototype demonstrated in a plant environment	This represents a major step up from TRL 6, requiring demonstration of an actual system prototype in a relevant environment. Examples include testing full-scale prototype in the field with a range of simulants. Final design is virtually complete.
8	Actual system completed and qualified through test and demonstration in a plant environment	The technology has been proven to work in its final form and under expected conditions. In almost all cases, this TRL represents the end of true system development. Examples include developmental testing and evaluation of the system within a plant/CCS operation.
9	Actual system operated over the full range of expected conditions	The technology is in its final form and operated under the full range of operating conditions. Examples include using the actual system with the full range of plant/CCS operations.

APPENDIX B: ACTIVE CARBON CAPTURE PROJECTS

(AS OF OCTOBER 2012)

Table B-1. **Post-Combustion Capture Projects**

Agreement Number	Performer	Project Title	TRL	Relevancy Statement
		Key Technology—Solvents		
FC26-07NT43091	University of Notre Dame	Ionic Liquids: Breakthrough Absorption Technology for Post-Combustion CO_2 Capture	3	Develop a new ionic liquid absorbent and accompanying process that overcome viscosity and capacity issues impacting cost and performance of ionic liquids by via "proof-of-concept" exploration and laboratory-/bench-scale testing of a variety of ionic liquid formulations.
NT0005498	University of Illinois	Development and Evaluation of a Novel Integrated Vacuum Carbonate Adsorption Process	3	Develop an integrated vacuum carbonate absorption process to improve absorption kinetics and lower regeneration costs by evaluating process parameters, identifying an absorption rate acceleration catalyst, and developing an additive for reducing regeneration energy.
ED33EE	Lawrence Berkeley National Laboratory	Development of Chemical Additives for CO_2 Capture Cost Reduction	3	Develop a solvent system that integrates amine, potassium carbonate, and ammonium solvents to enhance solvent absorption and reduce regeneration cost through bench-scale investigation of novel solvents.
FE0004274	3H Company, LLC	Post-Combustion CO_2 Capture for Existing PC Boilers by Self-Concentrating Amine Absorbent	3	Evaluate the feasibility of a self-concentrating absorbent capture process to determine capture costs and energy savings generated through use of an innovative material and process by developing an engineering design supported by laboratory data and economic justification.
FE0004228	Akermin, Inc.	Advanced Low Energy Enzyme Catalyzed Solvent for CO_2 Capture	3	Demonstrate the performance of an advanced carbonic anhydrase-enzyme-potassium carbonate solvent to improve sorption kinetics and decrease costs by conducing bench-scale testing to develop immobilized carbonic anhydrase enzymes to accelerate potassium carbonate uptake rates.
FE0005799	ION Engineering, LLC	Novel Solvent System for Post-Combustion CO_2 Capture	4	Develop an ionic liquid/amine mixture to realize cost and performance improvements through combination of two solvent systems by conducting bench-scale testing of an amine-based solvent with an ionic liquid instead of water as the physical solvent, greatly reducing the regeneration energy.
FE0004360	University of Illinois	Bench-Scale Development of a Hot Carbonate Absorption Process with Crystallization-Enabled High-Pressure Stripping for Post-Combustion CO_2 Capture	3	Evaluate the hot carbonate absorption process with crystallization-enabled, high-pressure stripping to determine process costs and technical feasibility by conducing lab-/bench-scale analyses of thermodynamic and kinetic data associated with major unit operations.
FE0005654	URS Group, Inc.	Evaluation of Concentrated Piperazine for CO_2 Capture from Coal-Fired Flue Gas	4	Investigate the use of aerosol formation in amine-based systems to decrease capture costs and energy use by conducting process analyses initially at a 0.1-MW scale and then scaled to 0.5 MW for testing at DOE's National Carbon Capture Center.
FE0007502	General Electric Company	Bench-Scale Silicone Process for Low-Cost CO_2 Capture	3	Enable a practical technology path for the use of a novel silicone solvent-based capture system that meets cost and performance goals via bench-scale analysis of process kinetic and mass transfer information and development of a manufacturing plan for the aminosilicone solvent.
FE0007466	Battelle Memorial Institute	CO_2 Binding Organic Liquids Gas Capture with Polarity-Swing-Assisted Regeneration	3	Develop a capture technology that couples nonaqueous, switchable organic solvents with a polarity-swing-assisted regeneration process to lower temperatures and energies for CO_2 separation by performing bench-scale analyses to determine process design parameters for eventual scaleup.
FE0007716	Babcock & Wilcox Power Generation Group, Inc.	Optimized Solvent for Energy-Efficient, Environmentally Friendly Capture of Carbon Dioxide at Coal-Fired Power Plants	3	Characterize and optimize the formulation of a novel solvent to lower capture costs by identifying blends that will improve overall solvent and system performance through bench-scale thermodynamic and kinetic analyses of concentrated piperazine blends with other organic compounds.
FE0007567	Carbon Capture Scientific, LLC	Development of a Novel Gas Pressurized Stripping-Based Technology for CO_2 Capture from Post-Combustion Flue Gases	2	Develop a novel gas pressurized stripping process to reduce CO_2 compression needs and the regeneration energy penalty through bench-scale tests of individual process units and computer simulations to predict the gas pressurized stripping column performance under different operating conditions.
FE0007741	Novozymes North America, Inc.	Low-Energy Solvents for Carbon Dioxide Capture Enabled by a Combination of Enzymes and Ultrasonics	3	Develop a capture system that combines a carbonic anhydrase enzyme with low-enthalpy solvents and novel ultrasonically enhanced regeneration to improve capture efficiency, economics, and sustainability by designing, building, and testing an integrated bench-scale system.
FE0007525	Southern Company Services, Inc.	Waste Heat Integration with Solvent Process for More Efficient CO_2 Removal from Coal-Fired Flue Gas	6	Develop a viable heat integration method to improve capture cost and performance by integrating a waste heat recovery technology (high-efficiency system) into an existing 25-MW pilot amine-based CO_2 capture process and evaluating improvements in energy performance.

Table B-1. **Post-Combustion Capture Projects**

Agreement Number	Performer	Project Title	TRL	Relevancy Statement
FE0007528	Neumann Systems Group, Inc.	Carbon Absorber Retrofit Equipment	4	Design, construct, and test an absorber that uses proven nozzle technology and an advanced solvent to reduce process equipment footprint and cost by conducting pilot-scale performance tests on 0.5-MW slipstream using a three-stage absorber unit and a best available technology CO_2 stripper unit.
FE0007453	Linde, Inc.	Slipstream Pilot-Scale Demonstration of a Novel Amine-Based Post-Combustion Process Technology for CO_2 Capture from Coal-Fired Power Plant Flue Gas	6	Refine a previously developed technology to reduce regeneration energy requirements by designing, building, and operating a 1-MW equivalent pilot plant using a novel amine-based solvent along with process and engineering innovations.
FE0007395	University of Kentucky	Application of a Heat-Integrated Post-Combustion CO_2 Capture System with Hitachi Advanced Solvent into an Existing Coal-Fired Power Plant	4	Develop a process using a two-stage stripping concept combined with an innovative heat integration method that utilizes waste heat to reduce costs through use of an improved power plant cooling tower by testing the process in a 0.7-MW slipstream pilot-scale system.
2012.01.05	National Energy Technology Laboratory	ORD Carbon Capture Field Work Proposal—Task 5: Post-Combustion Solvents	2	Develop technologies to capture 90% of the CO_2 produced by an existing coal-fired power plant with less than a 35% increase in to the COE as a critical step in reducing GHG emissions from fossil fuel-based processes by improving solvent working capacity, reducing sensible heat and heat of vaporization, and reducing the environmental impacts of solvent slip and degradation.
Key Technology—Sorbents				
NT0005578	SRI International	Development of Novel Carbon Sorbents for CO_2 Capture	5	Develop a novel carbon-based sorbent with moderate thermal regeneration requirements to evaluate the cost and performance capabilities of a low-cost sorbent via bench-scale parametric experiments involving fixed-bed adsorption and regeneration to determine optimum operating conditions.
NT0005497	TDA Research, Inc.	Low-Cost Sorbent for Capturing CO_2 Emissions Generated by Existing Coal-Fired Power Plants	4	Evaluate a low-cost alkalized alumina sorbent to determine the value of low-cost materials on capture cost and performance via bench-scale testing of a moving-bed capture system where adsorption and regeneration characteristics of the sorbent will be tested using actual flue gas.
FE0007804	Georgia Tech Research Corporation	Rapid-Temperature Swing Adsorption Using Polymeric/ Supported Amine Hollow Fiber Materials	3	Develop a rapid TSA process to evaluate cost and performance benefits of a novel hybrid capture approach via bench-scale testing of a module containing hollow fibers loaded with supported adsorbents surrounding an impermeable layer that allows for cooling and heating.
FE0007603	University of North Dakota	Evaluation of Carbon Dioxide Capture from Existing Coal-Fired Plants by Hybrid Sorption Using Solid Sorbents	3	Develop hybrid solid sorbent technology to decrease capture costs and energy use via bench-scale testing of a system that utilizes novel process chemistry, a low-cost method of heat management, and contactor conditions that minimize sorbent-CO_2 heat of reaction and promote fast CO_2 capture.
FE0007948	InnoSepra LLC	Novel Sorption-Based CO_2 Capture Process	3	Develop a sorption-based technology using a combination of novel microporous materials and process cycles to determine the impacts of this unique combination on capture costs and performance via bench-scale testing of system components using actual coal-based flue gas.
FE0007707	Research Triangle Institute	Bench-Scale Development of an Advanced Solid Sorbent-Based Carbon-Capture Process for Coal-Fired Power Plants	3	Develop an advanced process using molecular basket sorbents to evaluate the viability by developing fluidizable molecular basket sorbent production techniques, collecting critical process engineering data, and testing a continuous bench-scale molecular basket sorbent capture system using coal-fired flue gas.
FE0007639	W. R. Grace & Co	Bench-Scale Development and Testing of Rapid Pressure Swing Absorption for Carbon Dioxide Capture	3	Develop a rapid PSA process to evaluate concept cost and performance benefits by testing a bench-scale system using a low-cost, structured adsorbent with low pressure drop, high mass-transfer rates, high capacity, and high availability that will enable large feed throughputs.
FE0007580	TDA Research, Inc.	Low-Cost High-Capacity Regenerable Sorbent for Carbon Dioxide Capture from Existing Coal-Fired Power Plants	3	Develop a low-cost, high-capacity CO_2 adsorbent to demonstrate its technical and economic viability through sorbent evaluation and optimization, development of sorbent production techniques, and bench-scale testing of the process using actual flue gas.
FE0004343	ADA-Environmental Solutions, Inc.	Evaluation of Solid Sorbents as a Retrofit Technology for CO_2 Capture	5	Refine the conceptual design of a commercial solid sorbent-based, post-combustion CO_2 capture technology to facilitate future scaleup efforts through process modeling and pilot-scale testing using a 1-MW equivalent slipstream at an operating coal-fired power plant.

Table B-1. **Post-Combustion Capture Projects**

Agreement Number	Performer	Project Title	TRL	Relevancy Statement
FE0000493	Ramgen	Ramgen Supersonic Wave Compression and Engine Technology	4	Develop a supersonic shock wave compression technology to decrease carbon capture and storage costs and energy use through the design and testing of unique stationary power plant compressor products based upon aerospace shock wave compression theory.
FC26-05NT42650	Southwest Research Institute	Novel Concepts for the Compression of Large Volumes of Carbon Dioxide	5	Design a compression system that decreases power consumption and capital costs through the development of a semi-isothermal compression process with cooling via an internal cooling jacket or refrigeration to liquefy CO_2 so that its pressure can be increased using a pump, rather than a compressor.
NT0000749	Southern Company Services, Inc.	National Carbon Capture Center at the Power Systems Development Facility	5	Develop the capability to evaluate a broad range of capture technologies to facilitate scaleup of cost-effective technologies through testing of processes for pre-combustion CO_2 capture, post-combustion CO_2 capture, and oxy-combustion.
2012.01.06	National Energy Technology Laboratory	ORD Carbon Capture Field Work Proposal—Task 6: Post-Combustion Sorbents	2	Develop technologies to capture 90% of the CO_2 produced by an existing coal-fired power plant with less than a 10% increase in to the COE by improving sorbent CO_2 working capacity and hydrophobicity, decreasing heat capacity, and increasing chemical and mechanical stability.
		Key Technology—Membranes		
FE0004278	American Air Liquide, Inc.	CO_2 Capture by Subambient Membrane Operation	4	Develop a capture system using subambient temperature with a commercial hollow-fiber membrane to evaluate cost and performance impacts of a hybrid capture approach via bench-scale testing that demonstrates high selectivity/permeance and mechanical integrity and long-term operability at low temperatures.
FE0004787	Gas Technology Institute	Hybrid Membrane/Absorption Process for Post-Combustion CO_2 Capture	3	Develop a hybrid capture technology that combines solvent absorption and a hollow-fiber membrane to leverage capture cost and performance advantages of two different capture technologies through bench-scale testing on synthetic and actual flue gas to evaluate mass transfer and regeneration.
FE0007514	General Electric Company	High-Performance Thin-Film Composite-Hollow-Fiber Membranes for Post-Combustion Carbon Dioxide Capture	3	Develop high-performance thin-film polymer-composite hollow-fiber membranes to improve system performance via bench-scale testing to tune the properties of a novel phosphazene polymer and decrease costs through development of innovative fabrication techniques.
FE0007632	The Ohio State University Research Foundation	Novel Inorganic/Polymer Composite Membranes for CO_2 Capture	3	Develop a design and manufacturing process for new membranes to improve system performance through bench-scale testing of a membrane with a thin selective inorganic layer embedded in a polymer structure and decrease costs through development of a continuous manufacturing process.
FE0007531	William Marsh Rice University	Combined Pressure, Temperature Contrast, and Surfaced-Enhanced Separation of CO_2 for Post-Combustion Capture	3	Develop a novel gas absorption process to improve capture cost and efficiency through bench-scale testing of a combined absorber/stripper with a very high-surface-area ceramic foam gas-liquid contactor with basic and acidic functional groups for enhanced mass transfer.
FE0007553	Membrane Technology and Research, Inc.	Low-Pressure Membrane Contactors for Carbon Dioxide Capture	3	Develop a new type of membrane contactor (or mega-module) to decrease capture costs, energy use, and system footprint through bench-scale testing of a module with a membrane area that is 500 square meters, 20–25 times larger than that of current modules used for CO_2 capture.
FE0007634	FuelCell Energy, Inc.	Electrochemical Membrane for Carbon Dioxide Capture and Power Generation	3	Demonstrate the ability of an electrochemical membrane-based system (molten carbonate fuel cell) to simultaneously capture CO_2 and deliver additional electricity to the grid through bench-scale testing of an 11.7 m2-area electrochemical membrane system for CO_2 capture, purification, and compression.
FE0005795	Membrane Technology and Research, Inc.	Pilot Testing of a Membrane System for Post-Combustion CO_2 Capture	5	Scaleup a high-permeance membrane and process design to determine parameters for further scaleup and demonstration of the membrane-based system through small pilot-scale testing of a 1-MW equivalent capacity membrane skid at the National Carbon Capture Center.
12036	Idaho National Laboratory	Bench-Scale High-Performance Thin-Film-Composite Hollow-Fiber Membranes for Post-Combustion Carbon Dioxide Capture	3	Develop high-performance thin film polymer composite hollow-fiber membranes to improve system performance via bench-scale testing to tune the properties of a novel phosphazene polymer and decrease costs through development of innovative fabrication techniques.

Table B-1. **Post-Combustion Capture Projects**				
Agreement Number	Performer	Project Title	TRL	Relevancy Statement
2012.01.07	National Energy Technology Laboratory	ORD Carbon Capture Field Work Proposal—Task 7: Post-Combustion Membranes	2	Develop technologies to capture 90% of the CO_2 produced by an existing coal-fired power plant with less than a 10% increase in to the COE by increasing membrane selectivity and permeability, as well as overcoming the low-partial-pressure driving force for CO_2 associated with the process.
2012.01.08	National Energy Technology Laboratory	ORD Carbon Capture Field Work Proposal—Task 8: Oxygen Production	2	Develop technologies that overcome the energy penalties associated with conventional cryogenic separation and emerging ion-transport-membrane technologies and focusing on the development of novel approaches that yield high-purity oxygen.

Table B-2. **Pre-Combustion Capture Projects**				
Agreement Number	Performer	Project Title	TRL	Relevancy Statement
Key Technology—Solvents				
FE0000896	SRI International	CO_2 Capture from Integrated Gasification Combined Cycle Gas Streams Using the Ammonium Carbonate-Ammonium Bicarbonate Process	4	Develop a technology using a high-capacity, low-cost aqueous ammoniated solvent to meet cost and performance goals through bench-scale proof-of-concept testing followed by small pilot-scale testing using a slipstream of coal-derived syngas.
2012.01.02	National Energy Technology Laboratory	ORD Carbon Capture Field Work Proposal—Task 2: Pre-Combustion Solvents	2	Develop technologies with a pre-combustion programmatic goal to capture 90% of the CO_2 produced by an existing coal-fired power plant with less than a 10% increase in to the COE as a critical step in reducing GHG emissions from fossil fuel-based processes by developing solvents with increased CO_2 working capacity and hydrophobicity to prevent the absorption of water and promote CO_2 capture at temperatures consistent with those of gas cleanup technology.
Key Technology—Sorbents				
FE0000469	TDA Research, Inc.	A Low-Cost, High Capacity Regenerable Sorbent for Pre-Combustion CO_2 Capture	4	Develop a low-cost, high-capacity sorbent to demonstrate its technical and economic viability by optimizing chemical/physical properties, scaling up production, and conducting long-term testing with simulated syngas containing contaminants and eventually with actual syngas.
FE0000465	URS Group, Inc.	Evaluation of Dry Sorbent Technology for Pre-Combustion CO_2 Capture	3	Develop high-temperature/pressure/loading capacity sorbents that combine the water-gas-shift reaction with CO_2 removal to minimize energy efficiency impacts by combining process simulation modeling and bench-scale sorbent molecular and thermodynamic analyses.
FE0001323	New Jersey Institute of Technology	Pressure Swing Absorption Device and Process for Separating CO_2 from Shifted Syngas and Its Capture for Subsequent Storage	3	Develop a cyclic pressure-swing-adsorption-based process that produces purified hydrogen at high pressure and a highly purified CO_2 stream to enable economic evaluation for potential larger scale use through process/equipment development/testing and data analysis to facilitate scaleup.
2012.01.03	National Energy Technology Laboratory	ORD Carbon Capture Field Work Proposal—Task 3: Pre-Combustion Sorbents	2	Develop technologies to capture 90% of the CO_2 produced by an existing coal-fired power plant with less than a 10% increase in the COE as a critical step in reducing GHG emissions from fossil fuel-based processes by developing sorbents with improved CO_2 working capacity, increased hydrophobicity, low heat capacity, and increased chemical and mechanical stability at elevated temperatures consistent with those of gas cleaning technologies.

Table B-2. **Pre-Combustion Capture Projects**

Agreement Number	Performer	Project Title	TRL	Relevancy Statement
		Key Technology—Membranes		
FE-10-002	Los Alamos National Laboratory	High-Temperature Polymer-Based Membrane Systems for Pre-Combustion CO_2 Capture	3	Develop a polymer membrane technology that operates over a broad range of conditions to improve capture cost and performance through bench-scale testing of multiple structures, deployment platforms, and sealing technologies with high selectivity/permeability and chemical/mechanical stability.
FE0001322	University of Minnesota	Hydrogen Selective Exfoliated Zeolite Membranes	3	Develop a silica molecular-sieve membrane to decrease capture system costs by lowering fabrication costs and enhancing long-term stability through hydrothermal stability tests of exfoliated silicate powders and bench-scale membrane testing under shifted syngas conditions with simulated feed.
FE0001181	Pall Corporation	Designing and Validating Ternary Pd Alloys for Optimum Sulfur/Carbon Resistance	3	Develop an optimized Pd alloy that is tolerant to contaminants while retaining high hydrogen flux and selectivity to decrease costs and facilitate warm gas cleaning by employing a combinatorial material design approach for rapid, high-throughput screening of ternary alloys.
FE0000470	Arizona State University	Pre-Combustion Carbon Dioxide Capture by a New Dual-Phase Ceramic Carbonate by a New Dual-Phase Ceramic Carbonate Membrane Reactor	4	Develop a dual-phase ceramic-carbonate membrane to enable a one-step process for combined water-gas-shift/CO_2 separation with the potential to lower capture costs by synthesizing stable, high-permeance/selectivity membranes and fabricating tubular membranes/modules.
FE0000646	Gas Technology Institute	Pre-Combustion Carbon Capture by a Nanoporous, Superhydrophobic Membrane Contactor Process	4	Develop a gas/liquid membrane contactor concept to evaluate potential cost savings through laboratory and bench testing using pure gases, a simulated water-gas-shifted syngas stream, and a slipstream from a gasification-derived syngas.
2012.01.04	National Energy Technology Laboratory	ORD Carbon Capture Field Work Proposal—Task 4: Pre-Combustion Membranes	2	Develop technologies to capture 90% of the CO_2 produced by an existing coal-fired power plant with less than a 10% increase in the COE as a critical step in reducing GHG emissions from fossil fuel-based processes by developing membranes with increased permeability and selectivity toward CO_2 as well as increased mechanical stability and performance at high temperatures and pressures.

APPENDIX C: ADMINISTRATION AND DOE PRIORITIES, MISSION, GOALS, AND TARGETS

ADMINISTRATION PRIORITIES

Presidential Goal—Catalyze the timely, material, and efficient transformation of the nation's energy system and secure U.S. leadership in clean energy technologies

PRESIDENTIAL ENERGY TARGETS

- Reduce energy-related greenhouse gas emissions by 17 percent by 2020 and 83 percent by 2050, from a 2005 baseline.
- By 2035, 80 percent of America's electricity will come from clean energy sources.

DOE STRATEGIC PLAN—HIERARCHY OF RELEVANT MISSION, GOALS AND TARGETS

SECRETARIAL PRIORITIES

- *Clean, Secure Energy:* Develop and deploy clean, safe, low-carbon energy supplies.
- *Climate Change:* Provide science and technology inputs needed for global climate change negotiations; develop and deploy technology solutions domestically and global.

MISSION

The mission of the Department of Energy is to ensure America's security and prosperity by addressing its energy, environmental, and nuclear challenges through transformative science and technology solutions.

GOALS

- Catalyze the timely, material, and efficient transformation of the nation's energy system and secure U.S. leadership in clean energy technologies.
- Maintain a vibrant U.S. effort in science and engineering as a cornerstone of our economic prosperity, with clear leadership in strategic areas.

TARGETS

- Sustain a world leading technical work force
- Deploy the technologies we have
  - Demonstrate and deploy clean energy technologies
  - Enable prudent development of our natural resources
- Discover the new solutions the nation needs
  - Accelerate energy innovation through pre-competitive research and development
  - Facilitate technology transfer to industry
  - Establish technology test beds and demonstrations
  - Leverage partnerships to expand our impact
- Deliver new technologies to advance our mission
  - Lead computational sciences and high-performance computing

- Use Energy Frontier Research Centers where key scientific barriers to energy breakthroughs have been identified and we believe we can clear these roadblocks faster by linking together small groups of researchers across departments, schools and institutions
- Use ARPA-E, a new funding organization within the Department, to hunt for new technologies rather than the creation of new scientific knowledge or the incremental improvement of existing technologies

FOSSIL ENERGY RESEARCH AND DEVELOPMENT

MISSION

The mission of the Fossil Energy Research and Development program creates public benefits by increasing U.S. energy independence and enhancing economic and environmental security. The program carries out three primary activities: (1) managing and performing energy-related research that reduces market barriers to the environmentally sound use of fossil fuels; (2) partnering with industry and others to advance fossil energy technologies toward commercialization; and (3) supporting the development of information and policy options that benefit the public.

CLEAN COAL RESEARCH PROGRAM

MISSION

The CCRP will ensure the availability of near-zero atmospheric emissions, abundant, affordable, domestic energy to fuel economic prosperity, increase energy independence, and enhance environmental quality.

STRATEGIC GOAL

Catalyze the timely, material, and efficient transformation of the nation's energy systems and secure U.S. leadership in clean energy technologies.

STRATEGIC OBJECTIVES

- Deploy the technologies we have
- Discover the new solutions the nation needs
- Deliver new technologies to advance our mission

STRATEGY

- Accelerate energy innovation through pre-competitive research and development
- Demonstrate and deploy clean energy technologies
- Facilitate technology transfer to industry
- Establish technology test beds and demonstrations
- Leverage partnerships to expand our impact

THIS PAGE INTENTIONALLY LEFT BLANK

ABBREVIATIONS

ADA-ES	ADA-Environmental Solutions
ARPA-E	Advanced Research Projects Agency-Energy
°C	degrees Celsius
CCRP	Clean Coal Research Program
CCS	carbon capture and storage
CCSI	Carbon Capture Simulation Initiative
CO_2	carbon dioxide
COE	cost of electricity
Btu	British thermal unit
DARPA	Defense Advanced Research Projects Agency
DOE	Department of Energy
EOR	enhanced oil recovery
°F	degrees Fahrenheit
FE	Office of Fossil Energy
FOA	funding opportunity announcement
GHG	greenhouse gas
H_2	hydrogen
H_2O	water
HHV	higher heating value
IGCC	integrated gasification combined cycle
IMPACCT	Innovative Materials & Processes for Advanced Carbon Capture Technologies
MTR	Membrane Technology and Research, Inc.
MW	megawatt
MWe	megawatt electric
MWh	megawatt hour
N_2	nitrogen
NASA	National Aeronautics and Space Administration
NCCC	National Carbon Capture Center
NETL	National Energy Technology Laboratory
NO_x	nitrogen oxides
O&M	operating and maintenance
O_2	oxygen
ORD	Office of Research and Development
PBI	polybenzimidazole
PC	pulverized coal
PM	particulate matter
PSA	pressure swing adsorption
psia	pounds per square inch absolute
R&D	research and development
RD&D	research, development, and demonstration
RUA	Regional University Alliance
scfm	standard cubic feet per minute
SO_2	sulfur dioxide
SO_x	sulfur oxides
syngas	synthesis gas
T&S	transport and storage
tpd	tons per day
TRL	Technology Readiness Level
TSA	temperature swing adsorption
UNDEERC	University of North Dakota Energy and Environmental Research Center
WGS	water gas shift

www.ingramcontent.com/pod-product-compliance
Lightning Source LLC
Chambersburg PA
CBHW051050180526
45172CB00002B/584